LANDSCAPE INDICATORS

INDIKATSIONNYE GEOGRAFICHESKIE ISSLEDOVANIYA

ИНДИКАЦИОННЫЕ ГЕОГРАФИЧЕСКИЕ ИССЛЕДОВАНИЯ

LANDSCAPE INDICATORS

New Techniques in Geology and Geography

Edited by

A. G. Chikishev

Moscow Society of Naturalists
M. V. Lomonosov Moscow State University
Moscow, USSR

Translated from Russian by

J. Paul Fitzsimmons

Department of Geology
University of New Mexico
Albuquerque, New Mexico

 CONSULTANTS BUREAU • NEW YORK–LONDON • 1973

The original Russian text, the proceedings of a conference on New Methods of Geographical Indicator Research held May 21-22, 1968, in Moscow, published by Nauka Press in Moscow in 1970, has been corrected by the editor for the present edition. This translation is published under an agreement with Mezhdunarodnaya Kniga, the Soviet book export agency.

Library of Congress Catalog Card Number 72-88886
ISBN 0-306-10875-5
© 1973 Consultants Bureau, New York
A Division of Plenum Publishing Corporation
227 West 17th Street, New York, N. Y. 10011

United Kingdom edition published by Consultants Bureau, London
A Division of Plenum Publishing Company, Ltd.
Davis House (4th Floor), 8 Scrubs Lane, Harlesden, London,
NW 10 6 SE, England

Printed in the United States of America

PREFACE TO THE AMERICAN EDITION

Indicator investigations in the USSR have entered a new phase. Until very recently the principal means of indication was by definite plant communities or by particular relief forms. But, with development of the theory and practice of such investigations, it has become increasingly clear that indicators may be any visually observable features of a landscape. The basis of these views is the position that intralandscape relations are universal and continuous. From this has arisen landscape-indicator science as a special new branch of geographic investigation. The objectives of this discipline are the development of theory and practical methods of using the external aspect of the landscape as an indicator of components difficult to observe. Thus, at the present time, along with particular indicator investigations (geobotanical, geomorphic, and others), we should distinguish composite or complex landscape-indicator investigations.

A distinguishing feature of landscape-indicator geographical investigations is that the broad integrated approach to the solution of any particular problem permits us to set up as an object of indication not only statistical situations but definite processes also. This opens up new and extensive possibilities for using indicators in the study of present-day geological, geomorphological, and other natural processes, and also for using indicators to evaluate predictions of the indicated processes having different effects on the landscape.

The present collection of articles in some measure reflects this new level of indicator study. The collection has been prepared on the basis of material from a conference on new methods of geographic-indicator investigations, conducted by the Geographic Section of the Moscow Society of Naturalists in May 1968. The newest trend of investigation – indicator processes – has been discussed in the papers of A. G. Chikishev, M. G. Ilyushina, and N. N. Darchenkova. The first of the indicated articles is devoted to different aspects of using landscape-indicator investigations for one of the most important groups of geodynamic processes – karst development. The second deals with questions of prediction during reclamation. The third examines the possibility of indicating the natural evolution of sands in connection with the dynamics of their fixation. Some methods of quantitative treatment of field data from landscape-indicator studies have been described in the paper of E. A. Vostokova. Along with composite works on landscape indication, the collection includes works on different questions concerned with particular indicator investigations: geomorphological, hydrological, and others (the papers of A. I. Spiridonov, I. S. Gudilin, T. N. Moiseeva, N. A. Gvozdetskii, I. P. Chalaya, N. P. Matveev, and others). The remaining articles touch on narrower problems of method, associated with standardization, with indicator interpretation, with classification concepts used in indicator studies, and, lastly, with extrapolation of indicator patterns or rules.

In going deeply into the theoretical content of indicator investigations, we simultaneously expand the range of their practical application. In the present collection we may find examples of using indicator methods for highway investigations, prospecting for groundwater to irrigate pasture land, geocryological investigations, geologic mapping, and various other projects. The geographic range of indicator investigations is thus very broad: from cold tundras in the north to hot deserts in the south, and also including alpine districts.

A fundamental element of indicator investigations is the interpretation of air photos, and this is given particular attention in the present collection. Such investigations admit effective integration with other methods, in particular with geophysical, morphometric, and penetrative-sounding, as reflected in a number of articles. Thus, we find promise of creating complexes of faster methods of field studies, results of highly accurate instrumental determinations, which, obtained for individual points, may be extrapolated to distant areas.

The rapid evolution of indicator investigations and the abrupt transition from particular geobotanical and geomorphic indicator studies to composite or integrated landscape-indicator investigations attests to the fact that the new techniques conceal many unrealized possibilities. We have not yet sufficiently used indicator techniques in solving problems of conservation in nature, in studying recreational problems, in working on recultivation, and in surveying the field of medical geography. We should thus expect further progress in geographic-indicator investigations.

A. G. Chikishev

PREFACE

Investigations of landscape indicators represent one of the most important and promising trends in modern geography. These investigations deepen our understanding of intralandscape and interlandscape relations, on the one hand, and permit us to establish reliably and quickly, by external landscape components, decipient (concealed or deceptive) features of the landscape that are relatively inaccessible to direction observation, on the other. Investigations of indicators follows primarily geomorphologic and geobotanical trends, but gradually the range of indicators and indicated objects (indicatees) has been expanded, and in recent years these investigations have taken on an omnigeographic character.

Investigations of landscape or geographic indicators have wide application in the evaluation of soils for agricultural purposes and in engineering-geological, hydrogeological, geological, geomorphological, and geochemical studies for utilization of relatively uninhabitable regions. Generalization of experiences in indicator research and definition of the promise and future development of such research are therefore urgent.

The basis of the present collection is material from the Conference on New Methods of Geographical Indicator Research, conducted by the Geographic Section of the Moscow Society of Naturalists in May 1968 for the purpose of exchanging scientific information. The Conference revealed a great variety of methods of investigation, brought to light a series of new and promising trends, and also furnished support for introducing quantitative analytical methods in indicator research. Thus, a gradual transition from morphographic indicator interpretation of air photographs to morphometric interpretation was very clearly noted, and there also appeared a tendency toward working out objective methods of air-photo standardization of landscape indicators.

In contrast to the collection of the Geographic Section of the Moscow Society of Naturalists "Plant Indicators of Soils, Rocks, and Subsurface Waters" (Trudy MOIP, Vol. 8), published in 1964 by Nauka Press,* in which attention was directed chiefly toward geobotanical indicators, the present collection is devoted to the possibility of indicator use of a wide range of landscape components. The range includes geomorphological, hydrographic, geobotanical, complex landscape, and man-made indicators, and also ways of using aerial methods for landscape indications. Chief interest lies in the new trends of investigation: indicator hydrography, indicator karst investigations, and also possible indicators in large morphological—structural complexes and evaluation of conditions for transportation movement where roads are lacking in various relatively impassable regions. Considerable attention was also given to theories of geographic indicator research during the conference.

Although the trends of indicator research represented in the present collection are rather diverse, one can but notice that the different types of indicators are used in somewhat discon-

---

* English translation: Consultants Bureau, New York (1965).

nected fashion. Until now we have not achieved joint use of all physiognomic components of the landscape as indicators. This will become possible, obviously, only from more thorough investigations that will reveal more clearly the relations among the various landscape components. On this foundation we must raise a closely integrated structure of the various indicator trends. Only then may we speak of landscape indicators in a complete picture. Solution of all the many problems relating to this is a matter of the future, and we must hope that the present work will permit us to understand and solve problems more fully than we have been able to heretofore.

A. G. Chikishev

CONTENTS

INVESTIGATIONS OF LANDSCAPE INDICATORS

S. V. Viktorov, E. A. Vostokova, and A. G. Chikishev

One of the new trends in modern geography is the investigation of landscape indicators, using the entire combination of readily observed physiognomic components of the landscape for determining those components that are less accessible to direct observation (decipient). In its general plan, investigation of landscape indicators involves the recognition of indicators and an analysis of their reliability, on the one hand, and the practical use of the discovered indicator patterns, on the other, expressed chiefly in conducting indicator interpretation of air photographs and in comparing different indicator maps of prognostic character.

Development of investigations of landscape indicators is a new stage in the growth of the theory and practice of indication. Until recently individual components of the landscape, especially vegetation and relief, have been used as indicators. Botanical, geobotanical, and geomorphological indicator investigations thus appeared. Of these, botanical and geobotanical indicator investigations have been most widely conducted, being based on the possible use of vegetation and plant material as indicators of various conditions of the geographic environment (Viktorov et al., 1962; Chikishev, 1960, 1964). These investigations were developed more thoroughly in their methodological relations than in other directions. Indicator geomorphology has developed rapidly, revealing systematic relations between the landscape and geologic structure. The possibility of using landscape features for shedding light on geologic structure was pointed out in the early stages of development of geomorphology. It is interesting that one work published in 1802 was called "An Experiment in Physiognomy of the Earth, or the Art of Interpreting Internal Structure from the Earth's Surface." Consideration of the interrelations between landscape and geologic structure has permitted successful investigation of the geology and tectonics of extensive parts of the Soviet Union. Geomorphological indicator investigations are not restricted to the discovery of morphological–structural patterns. Physiognomic features of the surface are widely used as indicators of the origin, development, and present dynamics of the landscape.

Indicator hydrography and indicator zoology are acquiring ever greater significance among the new branches of indicator investigation. The picture of the drainage network, its density, the depth of stream dissection, and the character of the transverse and longitudinal valley profiles are important indicators of rocks, tectonic structures, and types and forms of landscape. Thus, dendritic or broadly branching drainage patterns indicate rocks of homogeneous composition and structure and plains-like relief, whereas a radial pattern is characteristic of domal uplifts, large closed basins, and local synclinal and anticlinal structures. Other drainage patterns, classified according to form and character of tributary distribution in the overall system of the basin, also have indicator significance. However, it would be premature to speak of a systematic and well-developed system of methods in this field of indicator investigation. The same should be said for indicator zoology, which in great measure has taken only the first steps, although these have been rather successful.

Landscape-indicator investigations represent a new form of indication, in some measure synthesizing the features examined above and using as indicators the whole complex of physiognomic components of the landscape, which gives it a number of substantial advantages, both in its thorough grounding and in the multiplicity of indicator predictions. However, we should note that landscape indication does not replace other types of indicator investigation and does not stand in the way of their development, since botanical, geobotanical, geomorphological, and other types of indicators completely preserve their value for certain features (objects, of indication).

Landscape-indicator investigations, being essentially complex, are used for the manifestation of definite landscape components, their combination, and predictions of natural processes (Viktorov, 1962). There is special significance in the study of natural complexes chiefly of low rank (facies, land tracts, groups of land tracts) and their structures in engineering-geological evaluation of a region, and also in searching for groundwater, determining the suitability of ground for agricultural utilization in various natural zones and highland belts, planning highways and railroads, determining transportation conditions over swamps, and in other partial and complex investigations.

For engineering-geological evaluations of regions underlain by carbonate and sulfate rocks, the extent and trend of karst processes are very important. Without considering them it would be impossible to solve several problems relating to the national economy in regard to utilization of a region. Landscape-indicator investigations of karst, based on a study of the correlation between individual components of natural complexes, permit us to indicate from physiognomic components of the landscape, with comparative rapidity and high reliability, the principal conditions of karst formation, to predict karst processes and phenomena, and to explain the various natural factors by the nature and distinctive features of karst development. The role of landscape-indicator investigations in regions underlain by crystalline and unconsolidated rocks is large, especially for planning construction work in permafrost.

In hydrogeological investigations, indicators of groundwater are relief variations, vegetation, and, more rarely, the character of surficial deposits. Vegetation is a direct indication of depth of occurrence and mineralization of groundwater, and relief and surficial deposits are used as indirect indicators. The landscape-indicator method permits one to define hydrogeological conditions of a region, to determine the zones of recharge and discharge of groundwater and the direction of drainage, and to predict the distribution of deep aquifers.

Landscape-indicator investigations are very important in the study of swamps as natural complexes, and also from the viewpoint of transportation through swamps. This new applied form of landscape-indicator investigation is very promising, since, in connection with the utilization of new lands, especially swampy lands of Siberia, it becomes necessary to effect long-distance transportation without roads across swamps and swampy tracts.

Interesting results are obtained in landscape-indicator investigations of individual landscape components in belts of different elevation. Although such work is just beginning, we have already discovered that plant material differs sharply within identical rocks at different elevations, and also within a single altitudinal belt in rocks of different types.

One of the characteristic features of landscape-indicator investigation is the broad possibility of using it for detecting and predicting natural processes. Indications of processes represent a field of geography as yet but feebly investigated. In the course of development of indicator geobotany, individual investigators, using the sensitivity of reactions of the plant cover to changes in conditions of the medium, have been rather successful in solving specific individual problems of process indication. But, on the whole, the problem remains unsolved. This is due to the fact that, for indicating processes and, even more, for prediction of processes that have not yet begun, it is impossible to restrict oneself by an analysis of a single physiognomic landscape component, even if it is very sensitive. It is necessary to analyze the interdependence and interactivities of those physiognomic and decipient components that give rise to the given process or are exposed to its effect in highest degree. Therefore, indicators of processes in their character and content are primarily landscape indicators.

Of the individual trends in landscape indication of natural processes, the most timely at the present time are indications of Quaternary and recent tectonic processes and of processes due to man's activity on the earth's surface. In indications of tectonic processes, several forms have been developed independently of each other. An extensive and rich fund of material has been collected on this subject by geomorphologists and geologists. Geobotanists have also conducted work along this line, not so extensively and not related to the other. Lastly, a number of investigations in regions of salt domes have shown a definite geochemical distinctiveness in the manifestation of recent tectonic activity. Landscape-indicator investigations aid in making close correlations of work on geomorphological, geobotanical, and geochemical indications and in explaining the specific constitution and structure of landscapes developing in tectonically active segments of various natural zones and regions. Thus far, we have made but very insignificant progress in this direction; its development is in the future.

The problem of studying the earth's crust as the environment of man's habitation, set up by Sidorenko (1967), requires as one of its conditions the successful development of methods that may be used in predicting results of those directed and incidental processes due to the activity of man in nature. Special attention should be given indications of the second group of processes, which arise independently of the will of man, frequently destroying the fruits of his creative, rational, and transforming activity. The greatest significance is found in indications of the early stages of these undesirable processes, because further development may be prevented in this case. This trend of investigation is extremely variable, and different branches have been worked out in varying degrees of completeness. For example, indications of open grazing land and indications of salinization and swamping during irrigation have been rather thoroughly studied. Concerning indications of geodynamic slope processes due to man, there is an urgent necessity of synthesis on the basis of landscape studies relative to indicator data collected by geomorphologists, geologists, geobotanists, and specialists in other branches of knowledge.

Indications during reclamation work are very important. Here we should develop definitions both of stages of removal of toxic chemical compounds from reclaimed dumps as well as of soil-forming stages during reclamation.

The wide use of heavy equipment in transportation, which, when this is not on roads, destroys the surface layers of the soil, is common in sparsely settled regions (deserts, tundras) during prospecting work for oil, gas, and other mineral resources, and leads to dire consequences. The drifting of sand, conversion of gypseous soils into wind-blown hummocks, destruction of extensive areas of natural pasturage, and other phenomena of soil degradation appearing because of the indicated activity might have been prevented had the landscape indicators of these processes been made apparent, permitting the processes to be recognized and restricted. It seems probable that in all such investigations man-made indicators must be widely used, but the methods of their application is but little understood as yet.

An important prerequisite in working out practical methods for indicators of natural processes is solution of several theoretical problems. One of the most important of these involves structural aspects of landscape units, marking the foci of a process and the course of its development. It is obvious that we may here successfully use the concept of biocenotic horizons and ideas of lateral and radial geochemical and energy flux as developed by Byallovich (1960). Indications of such flux are probably of great aid in investigating areas where a particular process is at work. Thus, far this has been worked out only in small measure for geochemical dispersion of salts (Kovda et al., 1954; Vyshivkin, 1960).

It is also of interest to study to what extent we may transfer to landscape indication the views of Dokhman (1936) concerning ecological series of plants. Among these, a special group of ecological–genetic series is distinguished, in which the sequential distribution of substances in space reflects the change in time. If these views should be extended to other spatial series of landscape units, the possibility of landscape indicators of processes would be considerably expanded.

Geographic extrapolation of observed indicators is very important for a theory of process indicators. The essence of this view is that certain regions are characterized by a sequence of changes in physiognomic components of the landscape with any particular process already studied. It might be possible to speed up the indications considerably if the evolved series is extended to a region more or less similar to that where the indicator series was first established. The possibility of such extrapolation has not yet been sufficiently studied, however.

Extension of the number of indicators, expansion of their range, and consideration of indicator objects, not only under static conditions but in reference to their development, make it necessary to obtain wide use of objective and precise methods of recognizing and evaluating landscape indicators under different natural conditions. Although a rather large number of different reliability scales have appeared recently, chiefly of geobotanical indicators, they are still in some measure conditional and subjective (Viktorov et al., 1962; Vinogradov, 1964).

For facilitating subsequent analysis of the investigated system "indicators–indicator objects," it is necessary to standardize methods of recording the primary material of indicator investigations. During geobotanical indicator investigations, for example, special forms have long been used for this purpose. On these forms the results of combined observations of indicators and indicator objects in the standard (key) districts are recorded. In recent years, with the introduction in research work of cards with marginal perforations (for hand sorting) in landscape-indicator investigations, attempts have been made to use punch cards for recording observations directly in the field. Meanwhile, as experience has shown, it has proved most suitable to place descriptions of physiognomic components of the landscape jointly with the investigated indicator object on standard description forms, accompanying this procedure with official records in a journal, and to transfer the already prepared full descriptions of indicator and indicator object to punch cards (Vostokova, 1967). With such a system and with well-considered coding, a working card index of indicators will be created, by means of which a rapid selection of necessary primary information may be made (Vostokova and Vyshivkin, 1966).

The use of forms or punch cards for descriptions of primary information (raw data) favors the introduction of more objective methods of recording observations. In indicator investigations there is also wide usage of quantitative characteristics of indicators such as the transect method, specified areas such as Raunkiaer's classes of life forms for individual plant indicators, and so forth. Meanwhile, in geobotanical indicator studies, for example, to designate plant abundance in a sample area of the standard area we have used the subjective evaluation utilized by Drude, and the degree of plant cover is determined by eye. The time has come for a more precise quantitative description of the elements of relief and plant cover in standard areas, especially when they are the leading indicators of the investigated indicator objects. Such material may be considered perfectly satisfactory for subsequent analysis of the investigated system of indicator–indicated objects.

The importance of a quantitative description of indicators in gathering raw data is obvious. Quantitative descriptions are necessary for objective evaluation of the closeness of correlations, for the wide introduction of mathematical statistics in treating raw data, and for recognizing and evaluating indicators.

Until now, empirical comparison of linked observations of indicator and indicated object has prevailed in the recognition and evaluation of indicators. Most reliability scales for indicators have been constructed on the basis of analysis of such linking. The very nature of indicator investigations, however, assumes the necessity of wide introduction in indicator studies of objective methods for recognition and evaluation of the indicators. Such objective methods must include mathematical statistics, methods of recognizing and evaluating interrelations between two (or several) landscape components, one of which is readily observable.

During hydrologic indicator investigations, for example, the simplest methods have been used: computed correlation coefficients, construction of empirical lines of regression, and computation of the index of correlation (Vostokova, 1967). It is perfectly clear that these methods find wider application in indicator investigations since they give comparable material, and this permits use of more objective reliability scales in the evaluation of indicators. We should also point out that the simplest methods of mathematical statistics, even with small sampling, should preferably be more widely applied both in analyzing the indication range of any particular indicator and in studying the entire indicator spectrum of definite, indicated objects. Graphical methods (such as construction of point diagrams, distribution curves, etc.) and calculation of dispersion (standard deviation) may be used to describe the indication range of an indicator. The dispersion makes it possible to compare indicators of any particular indicated object, to distinguish the most reliable of the multiplicity of indicators.

It seems to us that the attempt to evaluate indicator methods by the amount of information is of definite interest (although the computation is made for an isolated point and not for regional investigations), since it marks a way of using information theory for evaluating information capacity of the methods, and, consequently, for comparing and evaluating individual indicators (Shemshurin, 1967).

The use of quantitative methods does not exhaust the first stage of indicator investigations, i.e., by the recognition and evaluation of the indicators. Quantitative analysis is also advisable for the practical use of recognized indicator patterns. This is also significant for indicator decoding.

LITERATURE CITED

Byallovich, Yu. P., Biogeocenotic Horizons, Trudy MOIP, Vol. 3 (1960).

Chikishev, A. G., "The relation of vegetation to soil and hydrogeological conditions on Chysovaya River terraces," in: Problems of Indicator Geobotany, Izd. MOIP, Moscow (1960).

Chikishev, A. G., "The relation of plant cover to climatic and soil–lithologic conditions in the Middle Urals," in: Plant Indicators of Soils, Rocks, and Groundwater, Trudy MOIP, Vol. 8, Nauka, Moscow (1964).

Dokhman, G. I., "An ecological–genetic classification of vegetation on the Isha forest-steppe," Byull. MOIP, Otd. Biol., Vol. 45, No. 3 (1936).

Kovla, V. A., Egorov, V. V., Morozov, A. T., and Lebedev, Yu. P., Patterns of Salt-Accumulation Processes in Deserts of the Aral–Caspian Lowland, Trudy Pochv. Inst. im. V. V. Dokuchaeva, Vol. 44 (1954).

Shemshurin, V. A., "Evaluation of the effectiveness of different methods of engineering-geological investigations by means of information theory," in: Material from a Seminar on the Use of Geophysical and Mathematical Methods in Hydrogeological and Engineering–Geological Investigations, Izd. VSEGINGEO, Moscow (1967).

Sidorenko, A. V., "The earth's crust and the activity of man," in: The Earth's Crust and the Activity of Man, No. 4, Moscow (1967).

Viktorov, S. V., "Indicator trend in modern geography," Byull. MOIP, Ord. Geol., Vol. 37, No. 6 (1962).

Viktorov, S. V., Vostokova, E. A., and Vyshivkin, D. D., Introduction to Indicator Geobotany, Izd. MGU, Moscow (1962).

Vinogradov, B. V., Plant Indicators and Their Use in the Study of Natural Resources, Vysshaya Shkola, Moscow (1964).

Vostokova, E. A., Theoretical Basis of Landscape–Hydrodynamic Investigations and the Method of Using Them in the Search for Groundwater in Deserts, Author's Abstract of her doctoral dissertation (Aftoreferat Diss. na Soisk. Uchenoi Step. D-ra Geogr. Nauk), Moscow (1967).

Vostokova, E. A., and Vyshivkin, D. D., "The use of punch cards with marginal perforations in building a card file of geobotanical descriptions for indicator purposes," Bot. Zhurn., No. 2 (1966).

Vyshivkin, D. D., "Geochemical landscapes of the Mangyshlak Peninsula," in: Problems of Indicator Geobotany, Izd. MOIP, Moscow (1960).

RECOGNITION AND EVALUATION OF INDICATORS
E. A. Vostokova

As a result of any geographic indicator investigation undertaken for solving particular problems in geology, engineering geology, hydrogeology, and so forth, a system of natural indicators should be established, indicators that possess physiognomic character and may reliably indicate the desired indicated object. The ultimate purpose of hydrologic indicator investigations is the establishment of hydrologic indicators, both direct and indirect, i.e., recognition of the nature of hydrologic indicator correlation, establishment of actual hydrologic indicator significance of the physiognomic landscape components, evaluation of the reliability of assumed hydrologic indicators, and determination of their possible extrapolation.

Despite the obvious importance of working out a proper and objective method for recognizing and evaluating the reliability of indicators, such matters have received little attention in the literature on indicator studies, and most of the work that has appeared contains merely the final results.

TABLE 1. Work on Recognition and Evaluation of Hydrologic Indicators

Stage	Basic work content	Ways of recognizing hydrologic indicators
Collection of raw data	Analysis of ecological-biological and geographical premises	Characteristics of indicated object
		Characteristics of indicator
	Joint observations of indicator and indicated object in nature	Description of standard areas
		Description of ecological profiles
		Description and mapping of key area
Treatment of raw data	Empirical comparison of linkage between indicator and indicated object	Explanation of hydroecologic range of indicator
		Explanation of range of indicators of a definite indicated object
		Empirical clarification of indicator value of indicators
	Statistical treatment of raw data	Comparison of hydroecologic ranges
		Determination of closeness of correlation between indicator and indicated object
Compilation of final documentation	Compilation of indicator schemes	Selection of indicators by evaluation of their reliability
	Compilation of legends for indicator cards	Selection of indicators with consideration of their significance and reliability

7

Recognition and evaluation of indicators are made on the basis of collected raw data on the interrelations between an indicated object and an assumed indicator. Some concept of this scheme as applied to hydrologic indicators may be obtained from Fig. 1.

In this paper, the simplest methods of treating raw data on recognition and evaluation of indicators are considered. The collection of these data and the compilation of the final documentation necessary for further use of the indicator data are not touched upon. The methods of recognition and evaluation of indicators have been tested in hydrologic indicator investigations in the deserts of Kazakhstan and Central Asia. In view of the simplicity of the methods, they are readily usable and require no special preparation. However, they may furnish investigators comparable material, based on more objective criteria than in the widely used "intuitive" method of recognizing and evaluating indicators.

Recently, interest in a more objective evaluation of indicators has been expressed by development of reliability scales (Viktorov et al., 1962; Chalidze, 1966). All these scales have been based on analysis of linkage between indicator and indicated object.

In practical work with these scales,* for the simplest analysis and evaluation of hydrologic indicators, "auxiliary tables of reliability" have been prepared. These are secondary working tables designed for recording the analytical results of standard descriptions. From these tables it is easy to establish indicator reliability of each indicator and to find the amplitude of the indicated object for which the investigated indicator is reliable.

Such "auxiliary tables of reliability" have been used for simplest analysis and evaluation of hydrologic indicators in particular types of desert.

The first method of analyzing standard descriptions involved selection of descriptions according to some definite feature. The presence of a card file on hydrologic indicators on punch cards (Vostokova and Vyshivkin, 1966) has greatly simplified and speeded up the selection of necessary standard descriptions producible by different indices.

When analyzing the facts of hydrologic indicator investigations, punch cards or forms with standard descriptions are grouped primarily according to type of desert. Within each type, punch cards with descriptions of salt and fresh groundwater indicators are distinguished. Standard descriptions of hydrologic landscape indicators of a definite type of desert are grouped on the basis of analyzing relief, first, and vegetation, second.

An auxiliary table has been prepared for recognizing reliable indicators of fresh groundwater. The first column of the table contains all hydrologic indicators in a given type of desert; the second records the total number of observations. The next two columns show the contingency frequency for fresh groundwater (absolute number and percent relative to total number of observations), and the last column indicates the evaluation of the indicator value of the investigated indicator according to the reliability scale. An example of determination of indicator significance is shown in Table 2.

During indicator investigations, it is very commonly necessary to consider the determinability of gradations among indicator objects, i.e., gradations that may not correspond to the ecological range of the indicators. For example, in evaluating drinking water in deserts, the following categories are frequently used: up to 1 g/liter, ultrafresh water; 1-3 g/liter, fresh water; 3-8 g/liter, brackish water; 8-12 g/liter, salty water; and higher than 12 g/liter, bitter-salty water. In this connection, when making indicator investigations, one must find indicators for such artificial categories.

_____
* The reliability scale of Viktorov, Vostokova, and Vyshivkin (1962) was used for evaluating the reliability of hydrologic indicators, with some modification of Vinogradov (1964) on numerical restrictions for evaluating reliability.

TABLE 2. Recognition of Reliable Indicators of Fresh Water about the Margin
of the Hummocky Sands of Kazakhstan

Indicator	Total number of associations	Frequency of linkage with fresh water		Evaluation of indicator
		Number of cases	% of total number of observations	
Margins of hummocky sands with communities of phreatophytes	92	73	79.3	Satisfactory
Including				
a) with groups of grass associations (Lasiagrostis)	32	28	87	Completely satisfactory
Of these				
Alhagi –Lasiagrostis	13	7	53.8	Doubtful
Lasiagrostis	4	3	75	Satisfactory
Sophora –Glycyrrhiza –Lasiagrostis	8	7	87.5	Completely satisfactory
b) with Artemisia associations	14	11	78.5	Satisfactory
c) with Alhagi associations	14	7	50	Doubtful

The group of landscape hydrologic indicators of fresh water within a particular type of desert is generally inhomogeneous for indicated depth of the groundwater. In order to find reliable indicators for certain groundwater depths, a supplementary analysis of this group was made by means of auxiliary tables. For this purpose, a table was constructed in which the first row indicated the depth intervals of occurrence of groundwater and the first column distinguished by species the plant community of actual hydrologic indicators. In the compartments of the table thus formed, the contingency frequency of each indicator with a specific depth interval of groundwater (fresh) was recorded. The contingency frequency of indicator with indicated object furnishes a basis for distinguishing the water-table depth for which the investigated hydrologic indicator is sufficiently reliable.

Thus, to find hydrologic indicators of sandy deserts from a card file of hydrologic indicators, containing 825 punch cards with standard descriptions, sandy deserts were sorted out (325 cards). Of these, 250 contained descriptions of direct indicators of fresh water. The indicators of such water had different units, and further grouping was therefore made according to relief and vegetation. First, the material was grouped according to the character of the relief and the ecological group of plants. Thus, from the total a definite group of indicators was sorted out (such as the group of descriptions of deflation basins with hummocky sand occupied by communities of glycophilic phreatophytes), and these were analyzed in detail by means of the auxiliary table (Table 3).

Similar operations have been carried out to refine the indicator significance of indicators for degree of mineralization of groundwater.

Thus, evaluation of indicator significance of hydrologic indicators by an auxiliary table involves, first, determination of the reliability of the scale adopted and, second, discrimination of sufficiently large categories for which the reliability of the indicator becomes maximal (Table 4).

As a result of comparison we may distinguish sufficiently reliable indicators of a definite range of depth of groundwater or of the mineralization of the groundwater.

TABLE 3. Indicator Value of a Group of Landscape Hydrologic Indicators: Deflation Basins in the Hummocky Sands of Kazakhstan

Indicator	Contingency frequency	Categories of depth to fresh groundwater						Total number of observations	Evaluation of indicator
		0.5-1.0	1.0-1.5	1.5-2.0	2.0-2.5	2.5-3.0	3.0-5.0		
Deflation basins in hummocky sands: a) with communities of glycophilic phreatophytes	With categories of depth	21	27	11	6	11	12	88	Absolute indicator of fresh groundwater at depths of 0.5-5.0 m; unreliable for more precise determination of depth to groundwater
	The same, % of total number of observations	23.8	31.8	12.3	6.6	12.3	13.2	100	
Including*: with groups and communities of Chosenia macrolepis, Elaeagnus angustifolia, and Scirpus lacustris	With categories of depth	11	13	3	1	1	1	30	Completely satisfactory for groundwater at depth of 0.5-2.0 m
	The same, % of total number of observations	36.6	43.1	10.1	3.4	3.4	3.4	100	
with Scirpus lacustris–Glycyrrhiza association	With categories of depth	3	8	2	1	—	1	15	Completely satisfactory for groundwater at a depth of 0.5-2.0 m; satisfactory for groundwater at a depth of 0.5-1.5 m
	The same, % of total number of observations	20.0	53.2	13.4	6.7	—	6.7	100	
with association of Isatis sabulosa and Melilotus officinalis	With categories of depth	1	2	2	2	2	6	15	Satisfactory for groundwater at 1.5-5.0 m; unreliable for more precise determination of depth to groundwater
	The same, % of total number of observations	6.7	13.4	13.4	13.4	13.4	39.7	100	

* The number of actual indicators in this group reaches 10. Only two examples have been cited to illustrate position in the method.

TABLE 4. Evaluation of Indicator Significance of an Indicator
(Deflation Basin in Hummocky Sands with Chosenia macrolepis,
Elaeagnus angustifolia, and Scirpus lacustris
in Western Kazakhstan)

Category	Category of indicated object (depth to groundwater, m)	Contingency frequency % of total number of observations	Evaluation of indicator
Initial	0.5-1.0	36.6	Doubtful
	1.0-1.5	43.1	
	1.5-2.0	10.1	Unsatisfactory
	2.0-2.5	3.4	No indicator for these depths
	2.5-3.0	3.4	
	3.0-3.5	3.4	
Generalized	0.5-1.5	79.7	Satisfactory
	0.5-2.0	89.8	Completely reliable

However, the very character of the patterns studied during indicator investigations and the considerable amount of information accumulated require the use of mathematical statistics for determining the accurate correlation between indicator and indicated object. Armand (1949) has considered it obligatory to use a mathematical test of correlation and of defining reliability.

In practical use of hydrologic indicator investigations, graphical statistical methods have been widely used (Rodman et al., 1960; Vostokova, 1961; Beideman, 1964). Mathematical statistics has not yet become a necessary link in indicator investigations, however, although a tremendous store of information has accumulated in making the investigations, permitting mathematical tests of correlation to be made between indicator and indicated object in a number of studies. In hydrologic indicator investigations these methods have been used for analyzing the character of the ecological range of landscape indicators, depth to water table, and mineralization of groundwater.

For recognizing the indicator significance of physiognomic components of the landscape, the descriptions of standard areas are generally analyzed. In hydrologic indicator investigations, assumed direct hydrologic indicators are analyzed and studied in detail, i.e., plant communities made up of phreatophytes. These are differentiated according to degree of reliability from preliminary examination of the literature and accumulated data on the investigated region.

Further development of hydrologic indicator studies has led to the use of an entire complex of physiognomic landscape components as landscape hydrologic indicators. In this development phreatophytes have by no means lost their value, but, on the contrary, have acquired even greater significance, since direct hydrologic indicators generally apply to zones of groundwater discharge, where the water is participating directly in the transpiration process. During field work, the proposed phreatophytes are usually investigated with special care, and great interest is manifested in objective analysis of the collected data.

In studying these or other phreatophytes or communities of such plants, hydroecological ranges are plotted by means of dotted diagrams (Fisher, 1958). Cards with descriptions of the investigated hydrologic indicator were sorted from the card file in order to construct such dotted graphs. The method of plotting graphs of hydroecological ranges is extremely simple: points for all observations of the investigated indicator are plotted on a coordinate system with depth to the water table on the horizontal axis and mineralization of the groundwater, in grams per liter on the vertical axis.

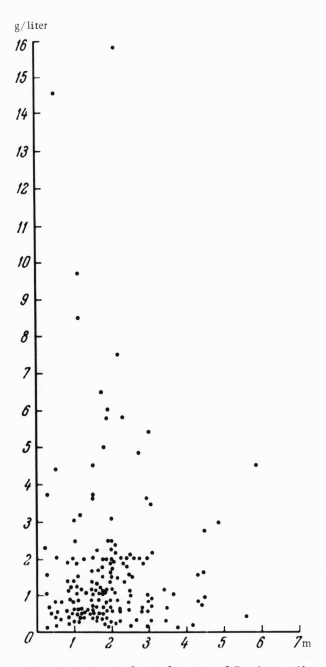

Fig. 1. Hydroecological range of <u>Lasiagrostis</u>
(dotted diagram).

On the resulting graphs, reflecting the magnitude of ecological range of a species or community relative to depth of water table and mineralization of groundwater, the coordinates of the points correspond to actual modes of occurrence and mineralization of groundwater at sites where the investigated plants grow. The extreme points designate extreme ecological conditions.

In Fig. 1 we show an example of a dotted diagram, reflecting the hydroecological range of <u>Lasiagrostis</u>.

Similar ecological ranges were plotted by Beideman (1964) in a study of the role of indicators in plant communities: direct hydrologic indicators on the Kura-Araks Lowland. Quantitative data used for plotting such dotted diagrams may serve for objective comparison of the investigated hydrologic indicators. Standard deviation has been used for comparative evaluation of the range of hydroecological amplitude of an indicator.

In indicator studies, standard deviation was first used by Viktorov to determine the distribution of an indicator on air photos according to the character of the indicated object.

In hydrologic indicator studies standard deviation was computed for comparison of hydroecologic ranges of the most widespread phreatophytes: Lasiagrostis splendens Trin, Phragmites communis Trin, Glycyrrhiza glabra L., Elaeagnus angustifolia L., Peganum harmala L., and others.

Standard deviation was calculated by the generally accepted method (Sapegin, 1937).

As an example we have shown the computation of standard deviation for two species of phreatophytes: the euphreatophyte Glycyrrhiza glabra L. (a) and the trychohydrophyte Peganum harmala L. (b). For simplicity of computation, it seems suitable to write all auxiliary calculations in tabular form immediately under the investigated series (Table 5).

$$\text{a) } n = \Sigma P = 90, \ M = \frac{\Sigma PW}{n} = 2.3, \ \Sigma P\alpha^2 = 235.71,$$

whence
$$\sigma = \sqrt{\frac{\Sigma P\alpha^2}{n-1}} = \sqrt{\frac{235.71}{90-1}} \approx 1.6 \ .$$

$$\text{b) } n = \Sigma P = 60, \ M = \frac{\Sigma PW}{n} = 4.0, \ \Sigma P\alpha^2 = 428,$$

whence
$$\sigma = \sqrt{\frac{\Sigma P\alpha^2}{n-1}} = \sqrt{\frac{428}{60-1}} \approx 2.7 \ .$$

The calculation of standard deviation (relative to groundwater depth) for phreatophytes differing in hydroecological amplitude has shown that, for euphreatophytes such as Phragmites communis Trin, Glycyrrhiza glabra L., and Lasiagrostis splendens Trin, standard deviation ranges from 0.6 to 2.5, whereas for trychohydrophytes (Peganum harmala L., Anabasis aphylla L., and several others) the range of standard deviation is from 2.6 to 3.5; i.e., the amplitude

TABLE 5. Calculation of Standard Deviation of
Ecological Ranges of Glycyrrhiza glabra L. (a)
and Peganum harmala L. (b)

	Classes of groundwater depth (W), m					
	1	3	5	7	9	10
a) P *	51	22	16	1	0	0
PW	51	66	80	7	0	0
$\alpha = W - M$	1.3	0.7	2.7	4.7	0	0
α^2	1.69	0.49	7.39	22.09	0	0
$P\alpha^2$	86.11	10.78	116.73	22.09	0	0
b) P	12	28	7	7	3	3

*Number in each class

relative to the investigated factor is much greater. Standard deviation serves as an objective index in comparing ecological ranges of individual hydrologic indicators and is thus a primary criterion of reliability of the investigated hydrologic indicators for certain categories of indicated objects.

For a complete description of the hydroecological range of any particular phreatophyte, standard deviation is also calculated relative to mineralization of the groundwater. Calculation of standard deviation has shown that species of broad ecological range relative to mineralization of groundwater have $\sigma \gtrless 4.5$, whereas for species of narrow ecological range $\sigma \lessgtr 1.0$. For example, for Elaeagnus in the northerm sandy desert, standard deviation relative to mineralization of groundwater is 0.57, whereas for Phragmites it is 4.9.

Having a diagram of the hydroecologic range of any particular hydrologic indicator and having the standard deviation relative to the investigated factors, the hydrogeologic range may be characterized succinctly by a conditional equation involving standard deviation, minimal, maximal, and modal values of factors, and the coordinates of extreme points. For example, the hydroecological range of Glycyrrhiza in the conditional equation has the following form:

$$A_{Gl.he} = \frac{1.6\sigma_{x\,max\,7.2}^{min\,0.5} \cdot M_0\,2}{0.5\sigma_{y\,max\,4.5}^{min\,0.1} \cdot M_0\,1}\,(x5;\,y4.5)\,(x7.2;\,y0.8),$$

where $A_{Gl.he}$ is the hydroecologic range of Glycyrrhiza; $\sigma_{x\,max7.2}^{min0.5}$ is standard deviation relative to groundwater depth at minimum (0.5 m) and maximum (7.2 m) level; $\sigma_{y\,max4.5}^{min0.1}$ is standard deviation relative to groundwater mineralization at minimum (0.1 g/liter) and maximum (4.5 g/liter) mineralization; the coefficient before the σ indicates the value of standard deviation; M_0 is the modal value of the factor, i.e., the groundwater depth or the mineralization most probable with the growth of Glycyrrhiza; and (x 5, y 4.5) are the coordinates of the extreme points.

The equation for the hydroecological range of an indicator with designation of the coordinates of extreme points furnishes a representation not only of the maximum value of some factor but also the interrelations between factors. We find reflection of the fact that the extreme conditions for one factor are compensated by more favorable conditions for the other. In the cited example, Glycyrrhiza grows where the water table is at considerable depth (7.2 m) and mineralization of the water is very slight (0.8 g/liter).

Equations of hydroecological ranges of indicators may be formulated not only for individual phreatophytes or communities of such phreatophytes but for any other hydrologic indicators (with consideration of their geographic and ecological aspects, of course).

With recognition of the indicator significance of the proposed indicators, great interest is manifested in studying the nature of the frequency distribution relative to the studied factor, i.e., to water-table depth or groundwater mineralization. On the basis of "auxiliary reliability tables," examined above, empirical series of regressions of the investigated indicators are plotted.

The resulting data may be represented on a contingency graph (empirical line of regression), ordinarily appearing as a broken curve.

To plot the contingency graph for hydrologic indicators, categories of the indicated object (groundwater mineralization or groundwater depth) are placed on the horizontal axis and the frequency of the contingency between hydrologic indicator and given object are represented on the vertical axis. Lines connecting points of these values also represent a contingency curve or an empirical line of regression.

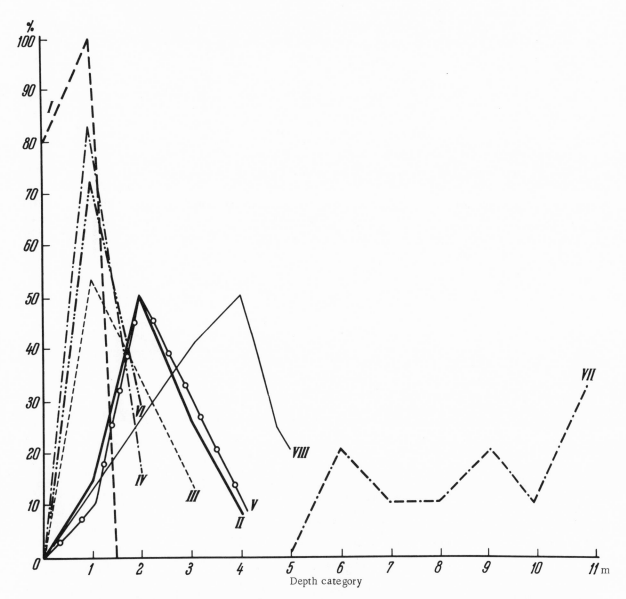

Fig. 2. Depth distribution of water table for different plant associations in Western Kazakhstan. I) Association with dominant Phragmites, Scirpus, and Typha; II) with dominant Lasiagrostis; III) with dominant Calamogrostis; IV) including Chosenia, Elaeagnus, and the psammophyte Scirpus; V) with dominant Elymus; VI) Halocnemum; VII) Artemisia—Anabasis; VIII) association with Tamarix.

By one form of empirical line of regression it is always possible to establish categories of indicated objects for which we must commonly find the investigated indicator. Thus, in Fig. 2 we show empirical lines of regression constructed for different plant communities in semiarid regions relative to depth of the water table. It is perfectly clear that for communities of Scirpus lacustris and Glycyrrhiza glabra the optimal depth of groundwater is 1.5-2.0 m, and for associations with Tamarix, the most favorable depth is 4.5-5.5 m.

Analogous methods have been used for studying indicators of mineralization of groundwater.

Greatest interest in indicator studies, however, is expressed in methods of quantitative evaluation of the closeness of correlation between indicator and indicated object. Such data may supply objective material for determining reliability of indicators.

During hydrologic indicator investigations, only cross correlation between indicators and groundwater has been established. Such relationship, taken by abstraction from other intra-landscape relations, has been called hydrologic-indicator correlation. To show the closeness of the correlation, the correlation coefficients and polychoric correlation index are computed (Rokitskii, 1961; Plokhinskii, 1961).

To find the correlation coefficient a correlation matrix was constructed by means of the previously discussed auxiliary tables. Thus, for computing the correlation between phreatophyte associations on flood plains of large and small streams on the one hand and fresh groundwater on the other, a correlation matrix was constructed (Table 6) in which computed correlation coefficients of such sites with fresh groundwater were added as an auxiliary in the column on the far right.

TABLE 6. Correlation Matrix for Computing Correlation Coefficient

Indicator	Frequency (P)		P_y	t_y	Correlation coefficient with fresh water
	Fresh ground-water	Saline ground-water			
Flood plains of large streams with phreatophyte communities	31	8	39	33.5	0.9
Flood plains of small streams with phreatophyte communities	10	19	29	33.5	-0.3
P_x^*	41	27			
t_x	33.3	33.3			

P_y^*) sum of values along horizontal
P_x) sum of values along vertical
$n = \Sigma P_x = \Sigma P_y$ — total number of observations = 68.
$$t_x = \sqrt{P_x(n - P_x)}; \quad t_y = \sqrt{P_y(n - P_y)}; \quad r = \frac{nP - P_x P_y}{t_x t_y}.$$

Substituting numerical values, we obtain r = 0.9 and r_2 = -0.3.

TABLE 7. Correlation Matrix for Calculating the Correlation Coefficient between Solonchak Margins and Fresh Groundwater

Indicator	Frequency (P)		P_y	t_y
	Saline ground-water	Fresh ground-water		
Margins of solonchak:				
Only with halophytes	35	1	36	29.4
Including growths of Lasiagrostis	2	7	9	21.3
Including Alhagi and others	12	3	15	25.5
	49	11		
t_x	23.1	23.1		

For reliability control, the least reliable r was determined:

$$\min r = \frac{\sqrt{n+36} - \sqrt{n}}{6} = 0.25.$$

$r_1 > \min r$, since $0.9 > 0.25$; $r_2 > \min r$, since $0.3 > 0.25$.

Consequently, both coefficients are reliable. But, whereas the correlation is positive in the first case (i.e., the indicator is a reliable marker of fresh water) it is negative in the second (the indicator does not point to fresh water).

To find the reliability of the correlation between the margins of solonchak soils (occupied by a hydrogenous hypsometric–hydrochemical series involving associations of Lasiagrostis) and fresh groundwater, the correlation coefficient is calculated by means of a correlation matrix (Table 7).

The correlation coefficient r = 0.65. To evaluate the reliability of this coefficient, the standard error was calculated:

$$m_r = \frac{1-r^2}{\sqrt{n}} \approx 0.07, \text{ whence: } \frac{r}{m_r} = \frac{0.65}{0.07} \approx 9 > 3.$$

The correlation is reliable, i.e., margins of solonchak with associations of Lasiagrostis in the ecological series may reliably indicate fresh groundwater.

Correlation coefficients were also computed for finding the correlation reliability between any particular physiognomic landscape components and shallow groundwater, in general, and actual depth categories, in particular. A correlation matrix was also constructed for this calculation.

For example, the correlation coefficient between hummocky sands with basins containing associations of Lasiagrostis, Elaeagnus, and Glycyrrhiza (Kazakhstan) and shallow groundwater is 0.75.

Determination of the restriction of basins to definite depth values of groundwater revealed that these indicators show reliable correlation with groundwater at 0.5-2.0 m (r = 0.5 ± 0.08, r/m_r = 0.5/0.08 = 6.25 > 3, i.e., the correlation is reliable).

TABLE 8. Calculation of the Polychoric Correlation Index

Indicator	Categories of groundwater depth, m												Sum of individual observations n_2
	0.5-1.5			1.51-2.5			2.51-3.5			>3.51			
	P	P^2	$\frac{P^2}{n^2}$	P	P^2	$\frac{P^2}{n^2}$	P	P^2	$\frac{P^2}{n^2}$	P	P^2	$\frac{P^2}{n^2}$	
Deflation basins in hummocky sands:													
With "tugai" associations	24	576	19.2	4	16	0.53	1	1	0.03	1	11	0.03	30
With "churot" associations	11	121	8.1	2	4	0.3	1	1	0.1	1	1	0.1	15
With glycophilic associations	13	169	3.9	11	121	2.6	9	81	1.9	10	100	2.4	43
n_1	48			17			11			12			
$\sum \frac{P^2}{n_2}$			31.2			3.43			2.0			2.53	
$\sum \frac{\frac{P^2}{n_2}}{n_1}$			0.56			0.25			0.2			0.4	

$r_1 = 3$, $r_2 = 4$, $\alpha = 1.45$, $n = \Sigma n_1 = \Sigma n_2 = 88$, $\rho = \dfrac{\alpha - 1}{\sqrt{(r_1 - 1)(r_2 - 1)}} \approx 0.18.$

The polychoric correlation index (Plokhinskii, 1961) was also used to determine correlation between groundwater (indicated object) and its indicators. This index may be used also for refining and checking the significance of any particular indicator, for determining the proper categories of indicated object from the viewpoint of obtaining the most reliable results.

A correlation matrix was also constructed to calculate the polychoric correlation index, and was used in all calculations. The "auxiliary reliability tables" considered above served as the basis for constructing this correlation grid. An example of computing the polychoric correlation index is shown in Table 8 (for three landscape hydrologic indicators considered in the framework of facies).

Calculation of the closeness and reliability of correlation between indicators and indicated objects by means of simple mathematical statistics permits objective evaluation and comparison of indicator possibilities of any particular indicator. Development of information theory and experience in using information measures for determining interrelations in geographic investigations leads us to suggest that it will be expedient to use this tool in a number of indicator investigations.

LITERATURE CITED

Armand, D. L., "Functional and correlative associations in physical geography," Izv. Vses. Geogr. Obshch., Vol. 81, No. 1 (1949).

Beideman, I. N., "Significance of coindicators in the indicator properties of plant associations," in: Plant Indicators of Soils, Rocks, and Mineral Deposits, Trudy MOIP, Vol. 8, Nauka, Moscow (1964).

Chalidze, F. N., Vegetation as an Indicator of the Mechanical State and Relative Age of Alluvial–Deltaic Deposits of Syr Darya, Author's abstract of his candidate's dissertation, Moscow (1966).

Fisher, R. A., Statistical Methods for Investigators, Gosstatizdat, Moscow (1958).

Plokhinskii, N. A., Biometrics, Izd. SO AN SSSR, Novosibirsk (1961).

Rodman, L. S., Levin, V. L., and Polikarpova, L. D., Quantitative Characteristics of Hydrologic-Indicator Significance of Vegetation in the Northwestern Caspian Region, Scientific Reports of Colleges, Biological Sciences, No. 3 (1960).

Rokitskii, P. F., Principles of Variation Statistics for Biologists, Minsk (1961).

Sapegin, A. A., Variation Statistics, Sel'khozgiz, Moscow (1937).

Urbakh, V. Yu., Mathematical Statistics for Biologists and Physicians, Izd. AN SSSR, Moscow (1963).

Viktorov, S. V., Vostokova, E. A., and Vyshivkin, D. D., Introduction to Indicator Geobotany, Izd. MGU, Moscow (1962).

Vinogradov, B. V., Plant Indicators and Their Use in Studying Natural Resources, Vysshaya Shkola, Moscow (1964).

Vostokova, E. A., Geobotanical Methods of Prospecting for Groundwater in Arid Parts of the Soviet Union, Gosgeoltekhizdat, Moscow (1961).

Vostokova, E. A., and Vyshivkin, D. D., "Use of punch cards with marginal perforations in preparing a card file of geobotanical descriptions for indicator purposes," Bot. Zhurn., No. 2 (1966).

THE PLACE OF LANDSCAPE-INDICATOR INVESTIGATIONS IN SPECIAL INTERPRETATION OF AIR PHOTOGRAPHS

N. G. Nesvetailova and A. V. Sadov

At the present time special interpretation of air photographs is a rather complex process, including a study not only of the external structure of the landscape, which is directly reflected in the air photographs, but also the internal structure. The manifestation of this latter is due to intralandscape relations by means of which external landscape features stand out as indicators of some internal features, hidden from direct observation. This is the reason for the great significance of indicator investigations, establishing a basis for special and, in particular, engineering-geological interpretation of air photos.

Indicator investigations have been rather extensively developed in recent time (Viktorov, 1966; and others). It is clear that we may already speak of the development of an independent branch of science, landscape-indicator science, which is the study of intralandscape relations for recognition of the designated indicators.

The principal objectives of such investigation are the discovery of indicator correlations between indicator and indicated objects (the problem of indication in the narrow sense and of standardization) and the study of changes in these relations in space (the problem of extrapolation) and also, in part, in time (the problem of prediction), due chiefly to changes in the indicated objects. Different trends in the nature of indicators are distinguished in indicator investigations: complex landscape, geobotanical, geomorphological, hydrographical, man-made and others, characterized by specific features due to distinctive aspects of the indicators and the nature of their relations to indicated objects. When we start from a broad treatment of landscape, we should then recognize that all possible trends of indicator investigations are based on the study of particular landscape components and are therefore linked to a single complex of landscape-indicator investigations, among which we may distinguish, to a great extent conditionally, groups of biological and geographic trends. Landscape principles permeate all types of indicator investigations and form the basis for interpreting indicator data, evaluating the spatial limits of applicability of indicator patterns, standardizing and selecting key areas, and so forth. The interrelation between individual indicator trends in the geographic group is especially well manifested in special interpretation, in which the most reliable results are obtained with integrated evaluation of the distinctive features of the region and indicators of different types are interrelated.

The term landscape-geological investigation, used by some authors, cannot be considered a fortunate one, since it does not reflect the essential indicator quality of these investigations, and the arbitrary discrimination of one landscape component is improper. The introduction of special concepts (landscape-indicator interpretation or the landscape-indicator method of interpretation) is also hardly advisable, since different methods of interpretation do not now exist, particularly for engineering-geological purposes.

The place and significance of landscape-indicator investigations in engineering-geological interpretation are very clearly defined by the distinct demarcation of indicators and interpreted features. The discord in treatment of interpreted features, not yet quieted, considerably confuses and renders difficult our understanding the interpretation process.

In agreement with the most widely established views, interpreted features are divided into direct, indirect, and complex or composite groups (Al'ter, 1966, 1967; Vinogradov, 1959, 1966; Tolchel'nikov, 1966). By direct features we mean such features of air photos as size, form, shading, and color indirectly reflecting the photometric and geometric properties of objects. Indirect features are individual landscape components and morphological elements correlative with the investigated object and characterizing its properties, or it represents logical categories reflecting exclusively the interrelations between external and internal components of the landscape reflected in the air photos (Raizer, 1958). They are manifested on air photos by means of direct features. Composite features include peculiarities of the natural regional complex in which the features of internal structure are indirectly reflected, displayed on air photos chiefly by the design and structure of its photographic image. Actually the indicated groups of features represent two fundamentally different categories, the first of which characterizes the peculiarities of the photographic image of the objects, and the other two represent composite categories, including indicators of investigated objects on the one hand and peculiarities of the photographic image on the other.

The basis of indirect and composite interpretive features is thus found in corresponding particular and composite (landscape) indicators. These concepts become considerably clearer when by interpretive features we mean solely features of the photographic image on an air photo due to different aspects of external structure of the landscape, including not only size, form, color, and shading, but also design and structure (Nesvetailova and Sadov, 1970). Indirect and composite features then correspond to the indicated types of indicators. The concept of interpretation in this connection acquires two aspects: particular, characterizing direct interpretation of external features of the landscape, and general, characterizing the study of natural features of a region by air photos as a whole, including direct interpretation on one hand and landscape-indicator interpretation of materials on the other. The latter may be effected directly when the object of study coincides with the object of direct interpretation and is reflected directly on the air photo, and it may be done indirectly when the interpretation process is accomplished by the following scheme: direct interpretation of features of the external structure of the landscape and then engineering-geological interpretation on the basis of landscape-indicator patterns. It appears improper to us to derive methods of interpretation based on the use of different types of indicators in the rank of independent methods, such as direct, indirect, composite, indicator, landscape, and others. These methods are based on a circumscribed group of methods, including the landscape-indicator method of recognizing objects of study by their characteristic indices and the method of direct interpretation of these indices by their interpretive features as a means of identification on the air photos.

The subdivision of interpretive media in indicator and interpretive features corresponds more to the essence of the interpretive process and makes it possible to represent the process in a clearer form. In contrast to the structural scheme of Komarov's interpretive process (Bondarik et al., 1967), the proposed scheme embraces the entire cycle, including both preliminary office and field study as well as final interpretation, and culminating in preparation of the final maps. In this respect it approaches the logical model of interpretation proposed by V. F. Rubakhin (Komarov et al., 1967), differing from it in somewhat different subdivision of operations into stages corresponding to interpretation practice.

The structural scheme of interpretation we have proposed provides for subdivision into four successive stages. The first three stages constitute preliminary office interpretation.

The first, preliminary, stage includes study of *a priori* material (literature, field notes, etc.) to evaluate supplementary interpretive criteria, the appearance of interpretive features, and the development of a hypothesis of landscape-indicator relations by logical conclusions and analogies. By supplementary interpretive criteria we mean factors and conditions not displayed on air photos but determined by peculiarities of both air-photo image and landscape-indicator relation. These criteria may be: a) technical (type of film or paper, aerial camera, height from which photographs were taken, and so forth); b) atmospheric-optical (illumination, season and time picture taken, smoke, etc.); c) landscape (natural climatic zones, phenological phases of plant development, geologic structures, seismic activity). The indicated category corresponds to the concept of interpretive criteria of S. P. Al'ter (1966), but it has wider significance.

The second stage – the study of air photographs (direct interpretation) – includes sequential study of the whole structure of air-photo images and their detailed analysis. The first operation is a study of the landscapes and their structures in relation to geologic structure of the region in order to develop a hypothesis of subdivision into landscape-typological regions, the selection on this basis of key districts, and the delineation of composite indicators.

Interpretation of this kind embraces the entire region of mapping, but it may be effected by various means and in variable degree according to the scale of mapping, the character of the natural conditions, and, in this connection, the overall character of the proposed indicators. It is generally carried out with a stereoscope, but in regions when the indicator role of relief features is insignificant, the interpretation from air photos is apparently limited. With small-scale mapping, stereoscopic interpretation may be apparently carried out in selective fashion only in the key districts. The indicated operation is not absolutely essential, and, for example, is not carried out for relatively small areas where natural conditions are highly uniform, or in some types of linear investigation where the traverses are short and the belt of investigation narrow.

Detailed study of the air-photo image is made by means of a stereoscope for developing a hypothesis of the make-up and character of the external components of the landscape. It is perfectly natural that this hypothesis must be partial to some extent, since it reflects the external make-up of the landscape only in the aspect of its use for indicator purposes.

A formally different operation takes place in subdividing a territory into photo-image areas of different character, corresponding to the individual elements of external landscape components. Actually the identification is carried out in almost parallel procedure.

The third stage – reprocessing the data – includes operations on landscape-indicator interpretation of material of direct observation for developing a hypothesis of an engineering-geological model of the terrain, based on landscape-indicator analysis with consideration of the general geologic aspects of the region and supplementary interpretive criteria. Such interpretation is impossible without sufficient and reliable *a priori* data, and, in this case, this is realized from field investigations. Operations on the study of external and internal peculiarities of the landscape thus prove to be separated in time. A combination of these operations into a single level of data processing is therefore inadvisable.

The fourth stage of evaluation of the general situation includes operations for control and completion of the hypotheses of external make-up of the landscape and an engineering-geological model of the terrain. It also includes investigation of deficient landscape-indicator relations. Office control operations are made by selective advisory decoding (interpretation) with consideration of adjacent regions by the investigators to shed light on the degree of subjectivity in the interpretation and also the quality of the preliminary maps of external landscape and engineering-geological structures. Field control also takes place selectively in key districts by field interpretation for testing hypotheses of landscape-indicator relations on which the interpretations are based and checking the natural basis of interpretive features. In parallel with the control operations themselves, deficient landscape-indicator relations are investigated (especially in the absence of a hypothesis of an engineering-geological model of the terrain), and "supplementary" engineering-geological data are collected, data that cannot be obtained directly by interpretation of air photos.

The overall evaluation of the situation includes a final thorough analysis of *a priori*, office, and field data, hypotheses of external landscape structure, landscape-indicator patterns, and an engineering-geological model of terrain, a final detailed interpretation of air photos, and extrapolation–interpolation operations. These latter involve the distribution of engineering-geological data obtained by direct study of key districts in regions not covered by direct field study. They are based on zonal–regional landscape-indicator patterns with consideration of three-dimensional changes in indicator relationships.

The result of this entire complex of work appears in final maps of engineering-geological conditions (actual engineering-geological model of terrain) that may serve as a basis for predicting changes in engineering-geological conditions with time and for indicating a different kind of special engineering-geological subdivision into districts with consideration of actual conditions and objectives of construction. The basis of prediction must be a study of changes of landscape interrelations with time.

Greatest interest is found in changes that appear as a result of the engineering activity of man and that to some extent becoming reflected in the external features of the landscape. Landscape analysis of air photos is therefore a promising method of prediction. It may be based on a comparative study of air photos made in different years and also on an analysis of the developmental history of the landscape from a study of typological series characterizing changes of landscape with time in light of probability–statistical patterns of changes in landscape interrelations. The prediction of changes of natural conditions by landscape methods is of very great value, but it has thus far received very little study.

Consideration of the interpretive process according to the above scheme thus clearly indicates its importance in landscape-indicator investigation. Special and, particularly, engineering-geological interpretation at the present level of practice cannot be made without landscape-indicator analysis of air photos.

LITERATURE CITED

Al'ter, S. P., Landscape Method of Interpreting Air Photographs, Nauka, Leningrad (1966).

Al'ter, S. P., "Methods of interpretation, interpretive features, and principal stages of the interpretive process," Izv. Vses. Geogr. Obshch., No. 4 (1967).

Bondarik, G. K., Komarov, I. S., and Ferronskii, V. I., Field Methods of Engineering-Geological Investigation, Nedra, Moscow (1967).

Komarov, I. S., Rubakhin, V. F., and Safronov, A. G., "Interpretation of air photographs as a recognition and information process," Transactions of the 9th All-Union Committee on Aerial Surveying, March 15-20, 1965 (Trudy IX Vses. Soveshch. po Aéros"emke 15-20 Marta 1965 g), Nauka, Leningrad (1967).

Nesvetailova, N. G., and Sadov, A. V., The Concept of Interpretive Features and the Content of the Interpretive Process in Engineering-Geological Investigations, Scientific Transactions of the Central Scientific-Research Institute of Communications (Sbornik Trudov TsNIISa), No. 25, Moscow (1967).

Raizer, P. Ya., "The general theory of interpretation of air photographs," Izv. Vysshikh Uchebnykh Zavedenii, Seriya Geodeziya i Aérofotos"emka, No. 6 (1958).

Tolchel'nikov, Yu. S., "Some terms used in the literature in interpretation of air photographs," Dokl. Inst. Geografii Sibiri i Dal'nego Vostoka, No. 11 (1966).

Viktorov, S. V., Use of Geographic-Indicator Investigations in Engineering Geology, Nedra, Moscow (1966).

Vinogradov, B. V., "Interpretation of vegetation of arid and semiarid zones," Trudy LAÉR AN SSSR, Vol. 7, Moscow (1959).

Vinogradov, B. V., Aerial Methods of Studying Vegetation of Arid Zones, Nauka, Moscow–Leningrad (1966).

MORPHOMETRICAL BASIS FOR SELECTING
TYPICAL AIR PHOTOGRAPHS IN
GEOGRAPHIC-INDICATOR INVESTIGATIONS

S. V. Viktorov

At the present time, when indicator methods have spread into the most diverse branches of science, an important source of information for investigators utilizing these methods has become the handbooks and albums containing typical air photos of landscape indicators. These air photos should give a view of the most particular, characteristic illustrations of indicators made by air photography. Of special significance are such albums and handbooks made in works of geographic extrapolation (Vinogradov, 1966; Vostokova, 1963), carried out for indicator purposes.

However, acquaintance with most published and unpublished material of such character convinces one that the selection of typical air photos depends entirely on experience and knowledge of compilers of the handbook and is done by eye, on the basis of a survey of several air photos with similar physicogeographic content. In this process, one frequently selects air photos not so much typical in actual fact as calling to the compiler's attention some special feature, sometimes an extraordinary aspect. Thus, handbooks may contain not average characteristic photos of landscape indicators but photos of more or less deviate types.

Typical air photos are commonly called standard, and the process of their selection is called standardization. In order that standardization not become artificial, depending on the individual abilities and peculiarities of the investigator, but become objective, it is necessary to use quantitative indices, and these must be obtained by more or less standard methods of measurement and computation. In other words, there must be a morphometrical basis. At the same time, when considering the broad framework of endeavors in standardization, the basis must be rather simple and clear, within easy reach of the investigator-interpretor with ordinary geographic or geologic preparation.

To create such a foundation, it is necessary first to consider normal black and white air photographs from a purely physiognomic point of view. In most cases we observe a field of view consisting of alternating segments of different configurations and different photographic tones. These segments may in some cases be structureless, of a single tone, or they may exhibit definite structure, i.e., they may consist in turn of smaller segments, which may have the dimensions of small patches, spots, short and narrow bands, and the like.

In the final analysis, air photos may be divided into several elementary segments or plots that cannot be further divided without going to a different scale. These segments or plots may be called physiognomic elements of air-photo images. They are the simplest indivisible components of the image structure.

Among the physiognomic elements of the air-photo image, we may distinguish a number of groups, being guided by different principles of classification. Thus, according to the nature of the distribution, these may be: a) background, i.e., filling some considerable part of an air photo as a continuous field, and b) diffuse, i.e., disconnected, as separate isolated segments, distributed throughout the above background. According to the uniqueness of the position on the photograph, the physiognomic elements may be subdivided into a) oriented, i.e., arranged along some definite structural lines of the air-photo image, and b) unoriented, i.e., lying with no apparent ordered arrangement. In configuration and variation of tone, the elements may be highly variable.

A series of physiognomic elements, similar to each other in external aspect, may generally be found on each photograph. This permits us to divide all physiognomic elements of a given photograph into physiognomic groups. Each group of physiognomic elements may be represented by a different number of plots: for background elements the number may be small, even restricted to one; for diffuse elements, the number may range widely, possibly reaching very large numbers. The totality of groups of physiognomic elements of a given air-photo image may be called the physiognomic composition of the photographic image. This concept is apparently similar to that called an assemblage by Nevyazhskii and Sadov (1967).

The elements of each physiognomic group occupy a definite part of the surface of the air photo. For background elements this part of the surface consists of one or two large plots; for diffuse elements it may embrace a large number of plots of different sizes. The total area occupied by any particular physiognomic group we define as the cover (a term borrowed from geobotany), independent of the number of plots composing it. Its value is expressed in percent of area of the photograph.

For diffuse elements of a photograph, it is still possible to keep in mind the density of elements, i.e., the number of elements of a given group per unit area. The density is the average value of a series of computations of the number of elements in a given group in a series of equal-area plots spread over the photograph. The values obtained by these computations in individual areas we call abundance (or frequency of occurrence). Density is thus the average abundance of diffuse elements in a given group. We may note that Nevyazhskii and Sadov, who followed the path of partial transfer of geobotanical and landscape terminology in their interpretation, call the number of elements in a given physiognomic group the frequency. This is not altogether proper, since frequency, i.e., the probability that an element will fall in a randomly selected area, depends not only on the number of elements in an area, i.e., abundance, but also on their distribution, as convincingly shown in phytocenology (Greig-Smith, 1964).

Lastly, we should note that not all physiognomic elements of a given air photo, especially small diffuse elements, are visible with identical clearness. Physiognomic elements of a particular group are very frequently found at the margin of discernibility, being very small or having indeterminate outlines. This permits us to divide physiognomic elements into a) indisputably discernible and b) doubtful. In further discussion we shall have in mind only the indisputably discernible elements. We may note, however, that discrimination of the two indicated groups of elements is of value only for a certain scale of the photograph. If the investigator finds it necessary for any reason to transfer a doubtful element to an indisputably discernible element, it is sufficient to enlarge the photo somewhat and the purpose will be achieved.

Solution of the problem of objective standardization on the basis of quantitative indices is possible only when air photos are considered the sums of physiognomic elements joined into groups by morphologic identity or close similarity. With this approach to the air-photo image, the search for typical (standard) photos reduces to a search for photographs in which the physiognomic composition and distribution of physiognomic elements should be modal (i.e., recurring most frequently) for a series of air-photo images of a given type of terrain. In other words, the problem arises of how to discover methods by which we may determine the mode of composition and the mode of distribution of structural elementary units of the air-photo image.

To solve the first problem – discovery of the mode of physiognomic composition of the air-photo image – we propose a method of physiognomic spectra of the landscape (Viktorov, 1966b). The physiognomic spectrum of a landscape is a diagram (we used a circular diagram, but Sadov (1966) recommended a histogram) reflecting the modal structure of the air-photo image of a given type of terrain, determined from a series of computations of the cover of different physiognomic groups during analysis of a series of air photos of a given type of terrain. The technique of making the computation will be described below.

After the values of cover of the physiognomic groups have been analyzed for individual air photos, the mode of each physiognomic group is found, and the physiognomic spectrum of discovery of a standard photo is plotted, and the structures of individual photos are compared empirically with the spectrum. We thus solve the problem of finding a standard of physiognomic composition of the air-photo image for a given type of terrain.

Discovery of a standard of distribution is a much more complex problem, the more so because, in the final analysis, it must be combined with the standard of composition; i.e., it is necessary to find a photo that is standard both for composition and for distribution. The complexity of the problem lies in the fact that, in contrast to the concept of composition, which has strict meaning (percentage determination of the cover of different physiognomic groups), the concept of distribution is rather indeterminate, involving the characteristic of "degree of complexity" of elementary distribution, the background distribution, diffuse distribution, and the characteristic spacing between elements (for diffuse elements), and evaluation of the degree of associativity of elements into groups as well as determination of orientation of the elements. To express all this in any single value, as in determining the physiognomic composition, is extremely difficult. It would seem that the simplest characteristic would be the abundance of elements of different physiognomic groups and the density computed from this (average abundance). But this characteristic is used only for diffuse elements; it has no meaning for background elements. This limits its value for analysis of air photos. For the same reason, it is necessary to reject such methods as construction of Dokhman–Vostokova curves and other methods suitable only for diffuse distribution of elements, also determination of variance according to Svedberg, since this is based on determination of abundance in individual areas. A unique value describing the distribution of both diffuse and background elements is frequency, and it should be given special attention in objective standardization.

The concept of frequency arose in geobotany (Raunkiaer, 1934). In the geobotanical sense, frequency may be determined as the probability of finding any particular species of plant (independent of abundance) in a randomly selected area. For purposes of interpretation, we have used this concept (Viktorov, 1966a) and have defined it as the probability of finding an element of any particular physiognomic group in an arbitrarily selected area. Determination of the frequencies of elements on an air photo is made by dividing the surface of the photo into a number of randomly selected equal-area plots and by determining the presence or absence in the areas of a definite physiognomic element. Since no computations of area are made and we note only the presence of the elements, this value may be determined for both diffuse and background components of the air-photo image. Frequency is expressed as percentage ratio of the number of areas in which the given element is observed to the total number of areas distinguished on the photo. Frequency thus differs fundamentally from abundance, with which it was confused by Nevyazhskii and Sadov (1967).

The background character of an element normally appears at once in a very high frequency percentage; diffuse elements exhibit much lower values. We should keep in mind, however, that, in determining frequency, the size of the plots or unit areas in which we seek to identify the presence of the investigated element is of very great significance. If these areas are very large, not only background but, to some extent, diffuse elements exhibit in this case the same high frequency. If the areas are too small, it is possible that no elements fall in them, not even background elements. The choice of size of the unit area should therefore be made so that the area be slightly larger than the smallest physiognomic element of the photo.

Frequency may be determined not in randomly scattered areas but in strings of areas, oriented along lines that crisscross the photo. In this case, the obtained value should be called the "frequency along a definite profile." This method of determining frequency was proposed by Dobbs (1939), and it has been described in detail in the literature of geobotany (Brown, 1957). If the lines along which the plots are arranged follow a systematic change in natural conditions (such as change in the elements of relief), the results of the determination may allow one to plot a curve of the change of any particular element in relation to the geomorphic conditions. All these investigations expand our views concerning the structure of the terrain and they may be very important, but they do not replace the basic operation: determination of the frequency for the entire investigated photograph.

In considering the great variety of distributions of diffuse elements, it is advisable to supplement the given frequency values with some other indices used especially for analysis of diffusion. These include the distribution curves of Dokhman–Vostokova and the abundance curves of Dobbs (in our modifications) and computation of variance (standard deviation). These methods have been described many times in the literature (Viktorov, 1955, 1966a, 1966b; Vostokova, 1961), and we shall therefore not pause here to describe them.

The complex of methods proposed by us for objective standardization thus reduces to the following: a) for selecting a photo with modal physiognomic composition, we construct a physiognomic spectrum and select a photo with composition most similar to the obtained spectrum, and b) for selecting a photo with modal distribution of elements, we determine the modal frequency of the most common elements and select a photo with frequency values similar to the modal, controlled by plotting frequency curves, Dokhman–Vostokova distribution curves, and also abundance curves.

Let us summarize the sequence and technical requirements of the standardization work.

1. The raw data for standardization must be a series of air photos assigned to the type of terrain for which a standard photo is to be found. The physicogeographic content of the air-photo image must be previously studied thoroughly during field indicator investigations till the contents of individual elements of the picture are determined.

2. All photos are divided into physiognomic elements. The divisions may be made either directly on the photo with India ink, outlining the boundaries of the elements, or on tracing-paper overlay.

3. The physiognomic elements are combined into groups. For each group of elements the cover is determined (by a planimeter or by an overlay of transparent plate with a millimeter grid.

4. For each photo a data card is prepared, on which are placed a general description of the terrain, the differentiated physiognomic groups, and the cover of each.

5. In comparing data cards of the photos, modal values of the cover are selected from the groups, and the physiognomic spectrum is plotted (though we may limit ourselves to preparation of a "modal data card," i.e., a table in which all physiognomic groups are represented

by their modal values). Physiognomic groups found on less than 30% of the analyzed photos may be ignored with no great loss to the precision of the work.

6. In comparing the data card thus prepared with the physiognomic spectrum (or with the "modal data card"), air photos are selected on which the physiognomic composition is nearest the modal. This photo will also be the standard of physiognomic composition. It is generally necessary to select not one standard, however, but several, since it is later necessary to seek standards of distribution from among them.

7. The frequencies of the principal elements are determined in searching for standards of distribution. The method of determination described in the literature (Viktorov, 1966a) involves moving an opaque plate with a rectangular window of definite size over the photograph (the size of the window is corrected according to the mean dimensions of the smallest investigated element; it is therefore convenient to have a series of windows of different sizes on the plate; it is necessary to use a window of a single size, however, for determining frequencies for all photos of a given type of terrain).

8. The frequency of elements of dominant physiognomic groups for individual photos is computed on the basis of the above determinations. By comparing the resulting data, the mode of the frequency is found. From the standards of composition, those are selected for which the frequency of the dominant physiognomic groups is nearest the mode.

9. If it is desirable to obtain supplementary proofs of the representative character of the selected air photos, one may construct Dokhman–Vostokova or Dobbs curves for elements having diffuse distribution, using methods described in the literature (Viktorov, 1966b).

LITERATURE CITED

Brown, D., Methods of Surveying and Measuring Vegetation, Commonwealth Agricultural Bureaux, Farnham Royal, Bucks, England (1954).

Dobbs, C. G., "The vegetation of Cape Napier, Spitzbergen," J. Ecol., No. 27 (1939).

Greig-Smith, P., Quantitative Plant Ecology, Butterworths, Washington–London (1964).

Nevyazhskii, I. I., and Sadov, A. V., Methods of Quantitative Evaluation by Air Photos of Morphological Characteristics of Natural Regional Complexes, Data of the Moscow Branch of the Geographical Society of the SSSR (Materialy Mosk. Fil. Geogr. Obshch. SSSR), Aérometody, No. 1 Mysl', (1967).

Raunkiaer, C., The Life Forms of Plants and Plant Geography, Clarendon Press, Oxford (1934).

Sadov, A. V., "Quantitative evaluation of the general structure of air-photo images of the landscape during engineering-geological interpretation," Byull. Nauchno-Tekhn. Informatsii Min-va Geologii SSSR, Seriya Gidrogeologiya i Inzhenernaya Geologiya, No. 1 (1966).

Viktorov, S. V., Use of the Geobotanical Method in Geological and Hydrogeological Investigations, Izd. AN SSSR, Moscow (1955).

Viktorov, S. V., Use of Indicator Geographical Investigations in Engineering Geology, Nedra, Moscow (1966a).

Viktorov, S. V., "Morphometrical analysis of air photos for indicating engineering-geological conditions," Byull. Nauchno-Tekhn. Informatsii Min-va Geologii SSSR, Seriya Gidrogeologiya i Inzhenernaya Geologiya, No. 1 (1966b).

Vinogradov, B. V., Aerial Methods of Studying Vegetation in Arid Regions, Nauka, Leningrad (1966).

Vostokova, E. A., Geobotanical Methods of Prospecting for Groundwater in Arid Regions of the Soviet Union, Gosgeoltekhizdat, Moscow (1961).

Vostokova, E. A., "Geobotanical methods of hydrogeological investigation in deserts," in: Groundwater Resources in Deserts and Their Uses, Izd. AN TurkSSR, Ashkhabad (1963).

THE SIGNIFICANCE OF ALTITUDINAL ZONES IN ROCK-INDICATOR INVESTIGATIONS IN MOUNTAINS

O. A. Osipova

Rock-indicator investigations in mountainous regions are among the most complex of such studies. Difficulties arise from the effect of altitudinal zones and exposure on the distribution of the plant cover, as a result of which the connection between rocks and outward aspect of a locality is masked, as it were, by climatic factors. It is possible that this explains why the number of indicator investigations in mountainous regions has been small. One of the first works devoted to special investigations in mountains is found in the paper of Chikishev (1960) on the Central Urals, in which the author has shown the potential of indicator methods for mapping the soil cover and subsoils in mountains. Later, questions of indicator studies in mountainous regions were considered by Lukicheva (1963) and Buks (1964).

More work in this branch of research is urgent, especially in the mountain massifs of Siberia, where indicator studies have been conducted to a very limited extent.

The work on which the present article is based was conducted in the Eastern Sayan, during the author's participation in the Central Geochemical Party of the Joint Thematic Expedition of the Irkutsk Geological Administration. Investigations were carried out in the Nizhneudinsk, Tulun, and Taishet regions of the Irkutsk Oblast. Both the highest mountainous part of the Eastern Sayan and the low-mountain and foothill flanks were studied, and the natural conditions in the region of work were therefore extremely diverse. In order to evaluate in full measure the variegation of sites of habitation encountered during the work, it appears proper to give a brief physicogeographical description of the areas investigated.

On the whole, the investigated region of the Eastern Sayan is divided by absolute height and relative variations into high-, medium-, and low-mountain districts.

In the high-mountain district, two morphogenetic types of relief are distinguished: high-mountain intensely dissected Alpine type and high-mountain dissected bald-top mountains.

Within the high-mountain intensely dissected Alpine type (1800-3000 m), characteristic geomorphic forms are: glacial valleys (troughs), cirques, horns, lakes in glacier-plucked basins, narrow sharp-crested ridges, and individual pyramidal peaks. The thickness of unconsolidated eluvial—deluvial deposits is irregular and negligible (0-0.5 m), but the slope-wash deposits at the feet of the slopes might reach thicknesses of several meters. Stream valleys, beginning here, are troughs with rather broad floors and channels very shallowly incised in these floors, and partly filled with glaciofluviatile deposits.

For the high-mountain dissected bald-tops (1700-2500 m), flat-topped divides and numerous individual flat crests with steep concave slopes are typical, exhibiting numerous traces of glacial activity such as glacial lakes and cirques. The thickness of residual material is small, ranging from 0.3 to 1.5 m. The slopes are covered by coarse blocks of talus material reaching considerable thickness (several meters).

In the medium-mountain district, morphogenetic types of relief are distinguished according to steepness of slope: medium-mountain steep-slope relief (1000-1700 m, slopes of 20-30°); medium-mountain moderate-slope relief (700-1600 m, slopes of 10-20°); and medium-mountain gentle-slope relief (600-900 m, slopes of 2-10°).

The formation of medium-mountain relief with different degrees of slope steepness has been clearly the result of varying intensity of erosional activity, due, in turn, to structural-tectonic features.

All these types of relief are spatially close to each other. Valley slopes are generally very steep, commonly of bedrock or covered by talus blocks. Near the divide crests the slopes gradually lessen, grow flat, grade into smoothed, rounded features of ancient erosion surfaces. On the divide surfaces and flattened slopes, weathering zones in the underlying rocks reach depths of 1-3 m.

The described features of relief are characteristic of areas in which igneous rocks and relatively flat-lying sedimentary rocks of Wendian, Cambrian, and Devonian age are present. Within areas of steeply dipping Archean and Proterozoic metamorphic rocks and, in places, igneous rocks, structurally controlled landscape has developed, steep ridges being present on the steeply inclined beds. Stream valleys are deeply incised, well developed, with terraces on both bedrock and alluvial fill.

The district of low-mountain gently sloping weakly dissected landscape is found at the junction of the Eastern Sayan foothills with the platform plains, the Kansk–Rybinskoe on the west and the Irkutsk–Chernikhovskii on the east, being the somewhat lower parts of a plateau. The elevations of these plains range from 400 to 650 m, locally reaching 750 m. This type of landscape is characterized by broad flat summit surfaces between streams, separated by shallow low-gradient valleys.

Below this district lie foothill aprons of fans and other alluvial material, grading into plains and terraces of alluvial and lacustrine accumulations. This type of landscape is of restricted occurrence, occupying a relatively small area in the foothill region of the Eastern Sayan. It is confined to Cenozoic foothill depressions. Its characteristic aspect is smooth topography, expressed in imperceptible transition from smooth, gently rolling divides with isolated ridges to very gentle slopes of the stream valleys. Quaternary deposits, here reaching considerable thickness (up to 100 m), form a significant part of this landscape.

Soils are equally varied (Karavaeva, 1958; Makeev, 1959). On soil maps of the investigated region, four types of soils are distinguished: high-mountain, mountain-taiga, taiga, floodplain-alluvial. A large number of subtypes are found in all four main types: nine in the mountain-taiga soils, five in taiga soils, four in high-mountain soils, and so on. In a number of places, the soil cover, depending on steepness and exposure of the slope, is almost mosaic.

A variety of physicogeographic conditions is responsible for a considerable variegation of the plant cover. Petrov (1948), Gluzdakov (1955), Smirnov (1958), Malyshev (1965), and others have studied the vegetation of the Eastern Sayan within the Irkutsk Oblast. All have noted a vertical zonation in the distribution of plants (mountain-taiga, high-mountain).

L. I. Nomokonovyi, M. V. Frolova, and G. A. Peshkova, workers at the Botanical Laboratory of the East Siberian Branch of the Academy of Sciences of the USSR (VSO AN SSSR), have prepared a vegetation map of the Irkutsk Oblast. On the basis of this map, two geobotanical maps have been published, on scales of 1:2,000,000 and 1:4,000,000, and a map of geobotanical regional subdivisions in the Atlas of the Irkutsk Oblast (1962). In the descriptions of the geobotanical regions, a list of formations is given and, for some places, associations. No descriptions of the latter have been provided, however.

In general, the Eastern Sayan is subdivided, in the region studied by the author, into two geobotanical districts: the Kazyr–Biryusa fir–cedar bald-top–taiga district and the Sayan–Khamar–Daban cedar bald-top–taiga district. The complexity and patchiness of the vegetation did not permit us, during our indicator investigations, to linger on such generalized divisions. Five subdistricts were therefore distinguished as a result of the work of the Central Geochemical Party: 1) high mountain–tundra brush–moss–lichen (in combination with subalpine and alpine meadows); 2) sub-bald-top–taiga larch–cedar and sparse larch–cedar, dwarf birch–ledum–rhododendron, and moss (with Rhododendron aureum); 3) medium-mountain taiga cedar–pine–larch variherbaceous–moss; 4) low-mountain taiga spruce–cedar–fir, variherbaceous–moss (with blueberry); and 5) piedmont–taiga larch–pine grass–variherbaceous. The first of the listed subdistricts is found chiefly at elevations of 1900-3000 m, the second at 1200-1900 m, the third at 800-1100 m, the fourth at 500-1000 m, and the fifth below 500 m.

In describing the relief, soils, and vegetation, it was rather easy to delineate at least three clearly defined zones in the mountainous part of the Eastern Sayan (not counting the foothills): the high-mountain tundra (with segments of meadows), the upper part of the medium-mountain forest (sub-bald-top–taiga), and the lower medium-mountain forest (medium-mountain taiga).

In the various summaries of the method of indicator investigation, it has been repeatedly emphasized that the development of any indicator scheme (including that for rock indicators) must be made relative to regions with more or less homogeneous physicogeographic conditions. Altitudinal zones must thus be assigned in application to the mountains of such regions. To test this view, we have analyzed standard descriptions of various rocks in the different zones. Analysis and comparison of the descriptions were made with dual purpose: to compare descriptions of a particular rock type (on slopes of similar steepness and exposure) but in different altitudinal zones, and to compare descriptions of different kinds of rocks in a particular altitudinal zone, where steepness and exposure of slopes are similar).

The area of distribution of any particular rock occurring in the various altitudinal zones has almost nothing in common with the physiognomic aspect or with the composition and character of the plant cover. This may be confirmed by numerous examples. In sandstones assigned by geologists to a particular formation (Zhaima), we find in the high-mountain zone (elevation of 1884 m) a dwarf birch forest-tundra with rare cedar, a carpet of lichen [Cladonia, Alectoria ochroleuca (Hoffm) Moss.] and blueberry (Vaccinium uliginosum L.), and in the upper part of the medium-mountain zone (at elevations of 1500-1600 m) occurs a cedar forest with larch and underbrush of honeysuckle (Lonicera coerulea L.), thickets of golden rhododendron (Rhododendron chrysanthum Pall.), and a carpet of moss. On diorites at an elevation of 1893 m grows a forest-tundra consisting of golden rhododendron, crowberry (Empetrum nigrum L.), and Bergenia crassifolia (L.) Fisch., but within 1 km, at elevations of about 1500 m, the same massif of diorites is covered with a birch–larch forest and a thick herbaceous cover of reedgrass [Calomagrostis arundinacea (L.) Roth.] and sedge. The number of such examples may be multiplied.

When we proceed to a comparison of districts on different rocks within a single zone, the picture changes sharply. We find a definite difference in the outward aspect, especially in the plants of comparable districts, which may have definite value as rock indicators. These differences are most clearly manifested for rocks of contrasting composition (such as between granites and limestones). On the Khan-Ude divide, for instance, at an elevation of 1800-2000 m, two districts 1.5 km apart have been described from slopes having a northern exposure and a steepness on the order of 20°. One of these districts, which we shall designate (1), is on granites, the other (2) on limestones. The district on granites is covered with typical tundra with golden rhododendron, birch (<u>Betula</u> <u>humilis</u> Schrank), crowberry, and willows. The herbaceous cover consists exclusively of sedges. The moss—lichen covers consist of <u>Cladonia</u> and species of the genera <u>Dicranum</u> and <u>Pleurozium</u>. In district (2) a forest-tundra has developed. The brush stage and brush and moss—lichen stage are very similar to the corresponding stages on the granites, but the herbaceous stage is very luxuriant, closely packed, having 10 different herbaceous species, including <u>Astragalus</u> <u>danicus</u> Retz and species of the genus <u>Trifolium</u>. Ferns are very abundant. The presence of trees and luxuriant herbaceous plants makes it easy to trace the boundary of the limestones.

Such comparisons of districts on granites and limestones in the upper part of the medium-mountain belt (at elevations of about 1400 m) were made on the crest near Gusharskoe Lake. In the wooded stage of district (3) (on granites), larch and cedar are dominant, and in district (4) (on limestones), larch and birch are also dominant, with an admixture of cedar. The difference is not great. Brushy growth also exhibited little differentiation, consisting chiefly of honey-suckle in both districts. The herbaceous cover shows strong contrast: on granites it is very thin, consisting of a suppressed cover of blueberry, but on limestones this cover contains up to 14 different species, the most abundant of which are <u>Calamagrostis</u> <u>arundinacea</u> (L.) Roth, <u>Veratrum</u> <u>Lobelianum</u> Bernh., and <u>Cypripedium</u> <u>guttatum</u> Sm., and it is very dense, closely packed, and high (especially because of the abundance of <u>Veratrum</u> <u>Lobelianum</u>).

Very noticeable differences were also noted in comparing districts on other rocks. In the high-mountain zone in the basin of the Khan River, for example, districts 1 km apart, one on granites, the other on gabbros, are covered by a forest-tundra with sparse larch and a rather uniform lichen cover. Comparison of the brush stage and herbaceous cover, however, shows that swampy shrubs (<u>Ledum</u> <u>palustre</u> L., <u>Empetrum</u> <u>nigrum</u>) and species of the genus <u>Carex</u> dominate on the granites, but rhododendron, variherbaceous plants, and ferns are dominant on the gabbros. Rather unexpected results, not yet fully explained, are obtained from a comparison of districts on the same rocks in the lower part of the medium-mountain zone on southern slopes in the basin of the Nizhigei River, where, though the other stages are similar, the herbaceous cover on the granites proves to contain many more species and to be more luxuriant than the cover on the gabbros.

The method described above has shown that it is possible to recognize also a number of other rocks in a single altitudinal zone: syenites, sandstones, diabases. The slightest differ-ence was noted between granites and gneisses.

In summary we may draw the following conclusions: a) in making rock-indicator studies, the first stage of the work should be delineation of altitudinal landscape zones with proper con-sideration of climatic, geomorphic, soil, and geobotanical data; b) rock-indicator schemes must be developed for individual altitudinal zones; and c) different stages of the plant cover have in-dependent rock-indicator significance. In the Eastern Sayan, the herbaceous cover possesses the most effective indicator qualities.

LITERATURE CITED

Buks, I. I., "Role of soil-forming rocks in the distribution of plant cover on the Olenek—Lena interfluve," Izv. Vses. Geogr. Obshch., Vol. 96, No. 2 (1964).

Chikishev, A. G., "Connection between vegetation and soil and hydrogeological conditions on terraces of the Chusovaya River," in Questions of Indicator Geobotany, Izd. MOIP, Moscow (1960).

Gluzdakov, S. I. "The vegetation of Tofalariya (Eastern Sayan)," Uchenye Zap. Novosibirsk. Gos. Ped. Inst., No. 10, No. 117 (1955).

Karavaeva, N. A., "High-mountain soils of Eastern Sayan," Pochvovedeniya, No. 4 (1958).

Lukicheva, A. N., Vegetation of Northwestern Yakutiya and Its Connection with the Geology of the Region, Izd. AN SSSR, Moscow–Leningrad (1963).

Makeev, O. V., Soddy Taiga Soils in the Southern Part of Central Siberia, Ulan-Udé (1959).

Malyshev, L. I., High-Mountain Flora of Eastern Sayan, Nauka, Moscow (1965).

Petrov, B. F., "Landscapes and soils of the central part of Eastern Sayan," Zemlevedeniye, Vol. 2, No. 42, Izd. Moskovsk. Gos. Univ. (1948).

Smirnov, A. V., "The forest in the upper reaches of the Uda River in Eastern Sayan," Trudy Vost.-Sib. Fil. AN SSSR, Seriya Biol., No. 7 (1958).

THE USE OF LANDSCAPE-INDICATOR METHODS
IN HYDROGEOLOGICAL INVESTIGATIONS
I. K. Abrosimov and Yu. M. Kleiner

The history of development of the indicator method in hydrogeology may be divided into at least three periods. In each one, hydrogeological problems were solved, and are solved, by using indicators differing in extent. The amount of information concerning hydrogeological phenomena and constants depends on this.

In the first and most significant period, in time, of development of the indicator method, which may be called the floristic, the indicator method was used to establish the presence of shallow groundwater and the mineralization of this water by individual species of plant indicators. It was natural that the indicator role of particular species of plant very commonly amounted to indication of some single characteristic of the groundwater. The indicator pointed to the presence or the depth of the water or, very rarely, to the degree of mineralization. The geobotanical period of indicator-method development in hydrogeology was associated with obtaining complex characteristics of groundwater. The use of plant communities as indicators of water permitted the construction of water-indicator maps of the first aquifer below the surface. These maps showed the depth of groundwater, the degree of mineralization, the geobotanical indicators, and they also displayed the results of sampling occurrences of water.

At the present time, the landscape-indicator method of prospecting for and mapping shallow groundwater is becoming ever more widely used in water-indicator investigation. The use of this method is based on the interrelation and interdependence of all landscape components, and it rests on our views concerning the laws of groundwater formation and dynamics. For hydrogeological purposes, the most useful indicators of groundwater among the physiognomic components of the landscape are relief, vegetation, and lithology of surficial rocks.

Vegetation in this technique is used as a direct indicator of depth and mineralization of shallow water. Relief and lithology of surface rocks become indirect indicators of water. Thus, in arid regions, slightly indurated eolian deposits are the sites of lenses of fresh water (Kunin, 1957; Ogil'vi and Chubarov, 1963), and these may be revealed by plant indicators in zones where such lenses discharge.

The use of complex physiognomic components for indicating groundwater permits us to determine not only the depth and mineralization of the water, but it also points us to the zones of formation and discharge of the water. This is of great practical significance, particularly for locating well sites.

The possibilities of using the landscape-indicator method in hydrogeology are restricted to shallow groundwater, chiefly local. In arid regions, the object of investigation is most commonly a lens or a stream of groundwater. Confined artesian water may be identified by the landscape-indicator method in zones of regional discharge, which may occur at tectonically

34

weakened sites. Such sites are most commonly due to faults. In arid regions water that rises along faults is frequently detected by the distribution of spring-formed hummocks and seeps. These hummocks are arranged along a line. Water that rises along a fault but does not pour out on the surface may be expressed on the landscape by lines of phreatophyte shrubs, lines along which small depressions may also occur. Recognition of deep groundwater by landscape indicators is thus recognition of indications of some manifestation of the water.

In all these stages of landscape-indicator studies, we are restricted practically to the very uppermost near-surface horizons of the earth's crust. Water indicators point to the first aquifer below the surface. Still, there can be no doubt that landscape analysis can tell us considerably more than this.

It is clear, in this connection, that this trend in water-indicator studies has developed under the influence of geobotanical indicator ideas; the relations between vegetation and landscape, and also between vegetation and groundwater, from the viewpoint of landscape indicators, has been studied much more thoroughly than the relation of landscape to geologic structure and lithology, which determine the conditions under which groundwater exists.

In recent theoretical works (Viktorov and Vostokova, 1966), lithology and tectonics have been considered along with mineral deposits as indicated objects, detected by means of indicators: plants and landscape. Without disputing this position on the whole, we wish to state that at the present level of geomorphological development, the position must be expanded and supplemented by means of morphologic-structural analysis. This is especially necessary in making regional hydrogeological subdivisions of any rather large territory.

The achievements of structural geomorphology, vigorously developed in recent years in connection with prospecting for oil, have permitted us to discover systematic relationships between landscape and geology for determining natural morphologic-structural regions.

Morphologic-structural analysis, determining the interrelations between geologic structures and geomorphic features, may be considered a component part of geomorphic-indicator investigation, which, in turn, is part of landscape-indicator research. The use of a number of physiognomic elements of the landscape as indicators of groundwater has become widespread among hydrogeologists. The complex interrelated use of these elements in landscape-indicator studies encounters a number of difficulties, however.

The use of a whole complex of landscape indicators in hydrogeology is predetermined by the very concept of landscape. Following Solntsev (1949), by landscape we mean a complex natural-regional assemblage in which the basis is chiefly geologic and genetic. The geologic-genetic basis of landscape is a component that determines all, or almost all, others.

Thanks to the achievements in the field of geobotanical and geomorphic indicators, and also to the intense development of methods of morphologic-structural analysis, this multiple approach is also applied to water-indicator studies. Consideration of all physiognomic elements of the landscape as indicators of water makes it possible to establish indicators not only for ordinary groundwater but also for interstratal water associated with certain geologic-genetic complexes, since these complexes are systematically reflected in definite morphologic structures. Such analysis leads us to make landscape-indicator investigation a necessary step into the subsurface, and also to predict the three-dimensional disposition of aquifers, which could be indicated by landscape–geobotanical methods only in districts with the most favorable landscape–geobotanical relations. Thus, sandy rocks at the crests of anticlines commonly accumulate meteoric water to form groundwater, which may be detected by geobotanical features: phreatophyte communities situated at the margins of the sandstone. The presence of groundwater in such anticlinal zones, with geobotanical features, permits us to make some *a priori* conclusions concerning the dynamics of groundwater and the conditions under which it occurs.

Multiple analysis — structural-morphological and landscape—geobotanical, combined into the more general "landscape-indicator" concept, permits us to study more thoroughly and profoundly the hydrogeological conditions of actual regions, both of the upper aquifer or lens of groundwater and also of deep-lying aquifers. The use of such analysis from the hydrogeological point of view helps us to solve a number of problems: prediction of deep groundwater, determination and delimitation of artesian basins, and determination of zones of groundwater recharge and direction of groundwater movement. These problems may be solved in different ways. For example, the boundaries of an artesian basin, which are now determined by analyzing the geologic structure and lithology of the region, may be refined by analysis of structural-morphological regions. Such regions for the most part determine also the mineralization of water within the artesian basin (Akhmedsafin, 1961).

Zones of groundwater recharge and direction of groundwater movement may be determined by the occurrence of certain morphologic structures, and also by certain forms and types of relief. Pronicheva and Zhernakov (1966), in particular, in the Kokdzhid sandstone mass (Primugodzhar'e), noted that positive forms of macrorelief and certain forms of microrelief (polygenetic barchans) are confined to the crestal zones of the uplift. Groundwater hence moves toward the limbs of the uplift.

Geologic processes now at work may be indices of definite structures and elements of these structures, and may thus indicate hydrogeological conditions. Karst development, landsliding, and deflation may be classed among such processes.

As an example we may examine a particular morphologic-structural region: the Ust-Urt and Southern Mangyshlak Plateaus. These plateaus occupy a broad region between the Caspian and Aral Seas. The surface is generally the bedding surface of limestone, and relief on the surface thus directly reflects tectonic features of the platform cover in the region. Thus, within the region of these plateaus, the largest orographic forms are clearly delineated first of all: broad, gentle, approximately east-west depressions in the northern and southern parts, restricted to the Northern Ust-Urt downwarp and the Southern Mangyshlak—Assake-Audan basin. Karst features are very characteristic of these zones (Kuznetsov, 1963).

Investigations have shown that both undoubted karst features and other features in which karst development was probably responsible for their early growth are practically absent in Northern Ust-Urt and over a broad area toward Mangyshlak and Southern Ust-Urt. This may be explained by difference in the hydrogeological regimes of the indicated regions. Structural-geologic studies in recent years have shown that the zone of the Southern Mangyshlak—Assake-Audan depressions is open toward the Caspian basin, but the Northern Ust-Urt basin is closed to the west. Thus, in the Northern Ust-Urt basin, there is probably no very appreciable flow of groundwater, the existence of which in the Southern Mangyshlak basin was suggested by Geller (1938) and has since been proved by a number of investigators (Mayantsev and Osyanin, 1965; Ulanov, 1965). According to computations of Ulanov (1965), based on pumping data from a number of holes drilled by the Mangyshlak Hydrogeological Party and the Hydrogeological Expedition of the Ministry of Geology and Conservation of Resources of the SSSR, groundwater discharge along the eastern shore of the middle Caspian, which corresponds in great part to the zone of the Southern Mangyshlak basin, amounts to more than 5.7 km^3/year.

It would seem that the view expressed above somewhat contradicts the broad development of karst in the region of the closed Barsa-Kel'mes basin. But, in our view, this fact not only agrees with the views expressed above, but it permits us to draw some very interesting conclusions, even if preliminary. Since the flow of groundwater is one of the most necessary conditions for development of karst, we can scarcely doubt that it is present. This confirms the view already expressed in the literature concerning seepage of the waters of the Aral Sea

(Blinov, 1956), the elevation of which is 52 m at its western shore, the eastern scarp of Ust-Urt. The most favorable structural and geomorphic conditions for this are found in the region of the Barsa-Kel'mes basin, and, seepage is apparently so appreciable that a stream of ground-water is formed, capable of moving the water at least into the Neogene rocks on the threshold that separates the Barsa-Kel'mes basin from the Southern Mangyshlak–Assake-Audan depression.

Thus, the distribution of karst features on the Ust-Urt and Southern Mangyshlak Plateaus is fully explained by the interrelated features of relief and tectonic structure of these regions that determine the character of the hydrogeology. The karst-forming processes are especially intense where, apart from the general conditions (the limestone cover and the stream of ground-water), particular conditions favoring karst are present: jointing in the limestone in the zone of local structures, large fractures or faults, and nearness to the sea. The sea washed across Ust-Urt in Late Pliocene time and extended along the erosional basins in the southern part of this region. This was noted first for the Assake-Audan basin by Fedorovich (1962).

Karst-forming processes are rather clearly indicated by vegetation. Viktorov (1955) noted that the initial stages of karst development may be detected by the nature of the plant cover, which differs sharply from the vegetation of surrounding areas. Sinks and canyon-like karst valleys, in addition to physiognomic expression in the landscape, are also marked by specific plant communities.

On the basis of morphologic-structural analysis, particularly on the distribution of karst features under definite specific and regional structural-geological conditions, it is possible to judge rather accurately the hydrogeology and tectonic features of broad regions. We should also note that within different morphologic-structural region, the nature and appearance in the landscape of such interrelations may be entirely different.

The literature furnishes many examples of the simultaneous relations of hydrogeological conditions to morphologic structures and vegetation. Rezvoi (1947) described an example of recent tectonic activity in Fergana, where patches of phreatophytes were found in gravels barren of vegetation. It was shown that the phreatophytes were confined to zones with shallow groundwater associated with the formation of hydraulic head from subchannel flow in connection with continuing uplift. Syrnev (1966) also noted similar occurrences of groundwater that could be detected by vegetation, springs, wells, and landslides toward synclinal depressions.

There is great interest in the multiple study of lineaments (straight segments of stream valleys, scarps, rectilinear shores, etc.), which are most commonly geomorphic expressions of deep fractures, very well reflected in the landscape because of the discharge of groundwater along these fractures. Inferred faults mapped during geological surveying emerge as indicators. But, since lineaments are not always associated with faults, geobotanical indicators may permit us to establish a hydrogeological anomaly, making it possible to clarify the relations of certain lineaments to faults. The expression of faults in the landscape is commonly more complex, however. Weakly fractured zones along faults may display not lineaments but chains of negative forms, the relationship of which with faults may be established only by their spatial arrangement. Such is the zone of karst occurrence along a fault on the northern limb of the Karabaur anticline in Ust-Urt.

The above discussion leads us to note some problems that must be solved for further improvement of methods of geographic-indicator investigations in hydrogeology. These are, first, regional subdivision of the landscape of extensive regions, especially plains, on a structural-geomorphic basis. The separation of natural structural-geomorphic regions characterized by common tectonic development, geologic structure and lithology, and uniform expression of these in the landscape and plant cover permits us to use known interrelations, and those that appear during the work, for proper interpolation of data obtained in separate key districts to the entire morphologic-structural region.

It is also necessary to make a special classification of relief forms that may serve as indicators for definite hydrogeological and engineering-geological conditions, such as karst and ravine forms, and also some man-made features, indicating the presence of water. We must emphasize here the necessity of establishing the age of the indicated forms and the relations between similar ancient and recent relief forms, taking into account the fact that they may prove to be (in case recent forms are lacking) indicators not of groundwater but of climatic conditions of the past, which is no less important but is of entirely different significance.

In addition, it is necessary to study those relief forms that possess indicator significance in different morphologic-structural conditions for systematic and thorough investigation of their relations to groundwater.

It is perfectly clear that the use of multiple landscape-geobotanical and structural-geomorphic methods appreciably enriches water-indicator studies and may prove to be of fundamental aid in hydrogeological investigations.

LITERATURE CITED

Akhmedsafin, U. M., A Method of Preparing Prediction Maps and Surveying Artesian Basins in Kazakhstan, Alma-Ata (1961).

Blinov, L. K., Hydrochemistry of the Aral Sea, Gidrometizdat, Moscow–Leningrad (1956).

Fedorovich, B. A., "Aspects of the migration of solutions and the formation of crusts and karst in deserts," in: General Questions in the Study of Karst, Izd. AN SSSR, Moscow (1962).

Geller, S. Yu., The Origin of Closed Basins, Trudy Inst. Geografii, Moscow (1938).

Kuznetsov, Yu. Ya., "The karst of Ust-Urt," Zemlevedeniye, Vol. 6, No. 46, Izd. Moskovsk. Gos. Univ., Moscow (1963).

Kunin, V. N., "Conditions of groundwater formation in deserts," Summary Reports of the 11th General Assembly of the International Geodetic and Geophysical Union, The International Association of Scientific Hydrology, Vol. 5, Moscow (1957).

Mayantsev, G. P., and Osyanin, Yu. A., "Groundwater discharge from Mangyshlak to the Caspian Sea," Okeanologiya, No. 5 (1965).

Ogil'vi, N. A., and Chubarov, V. N., "Study of the dynamics of moisture and the processes of its condensation in the zone of aeration," in: Lenses of Fresh Water in Deserts, Moscow (1963).

Pronicheva, M. V., and Zhernakov, P. I., "Description of objects of geomorphic investigation," in: Geomorphic Methods in Prospecting for Oil and Gas, Moscow (1966).

Rezvoi, D. P., "Traces of tectonic movements of the 'present day' in southern Fergana," in: A Collection on Theoretical and Applied Geology, No. 1, Geoltekhizdat, Moscow (1947).

Solntsev, N. A., "The morphology of natural geographic landscape," in: Questions of Geography (Vopr. Geografii), No. 16, Moscow (1949).

Syrnev, I. P., "Use of the multiple method of structural-geomorphic investigations in the southern Kara-Bogaz region," in: Geomorphic Methods in Prospecting for Oil and Gas, Moscow (1966).

Ulanov, Kh. K., "Groundwater discharge into the Caspian Sea and the migration of its water to the floor and the shore," Dokl. Akad. Nauk SSSR, Vol. 162, No. 1 (1965).

Viktorov, S. V., "Geobotanical features of karst-collapse processes in the desert," Byull. MOIP, Otd. Biol., Vol. 60, No. 1 (1955).

Viktorov, S. V., and Vostokova, E. A., "Landscape-indicator science and its value in geological investigations," Byull. Nauchno-tekhn. Informatsii Min-va Geologii SSSR, Seriya Gidrogeologiya i Inzhenernaya Geologiya, No. 1 (1966).

STUDY OF THE HYDROGEOLOGICAL CONDITIONS
OF ALLUVIAL FANS BY MULTIPLE LANDSCAPE
AND GEOPHYSICAL METHODS

N. N. Sharapanov and A. V. Shavyrina

A group of workers at the Geophysical Laboratory of the All-Union Institute of Hydrogeology and Engineering Geology are conducting experimental investigations on development
of a method of studying alluvial fans on the piedmont plains of Central Asia. This method is
very promising for large-scale hydrogeological mapping, since the combination of geophysical
and landscape methods may replace a great amount of drilling. Geophysical methods include
direct-current electrical surveying and induced potentials.

The proposed combination of methods of investigation may solve the following geological—
hydrogeological problems: 1) the grouping of alluvial fans by types according to morphology,
source of material, and possible degree of flooding; 2) determination of thickness and lithology
of the unconsolidated deposits in fans of each type; 3) study of the hydrogeological conditions
of these deposits (determination of water table, discrimination of zones of different mineralization, qualitative evaluation of reservoir properties of aquifers). On the basis of the data obtained, a hydrogeological map of the alluvial fans may be finally prepared.

Investigations were made on the piedmont plains framing the Fergana valley. They have
shown that, according to the character of the rocks, the depth of the water table, and some other
features of alluvial fans, the composition of multiple methods and the effectiveness of each method
are different. In this connection the first stage of investigation was the grouping of alluvial fans
in the investigated region according to conditions affecting the formation and supply of groundwater.

The following features represent the basis for distinguishing types of alluvial fans: the
presence or absence of a steady flow of water, and the amount of flow on which the composition
and thickness of the unconsolidated deposits depend in great measure; the slope of the surface
and the distance from the mountain mass (the source of the fragmental material). The latter
factor determines lithologic zonation, which involves accumulation of coarse fragments in the
upper part of the fan and a change downward to finer material. Lithologic changes are manifested in variations in expression of the hydrogeological zones: the zones of recharge, underground flow, and discharge. This separation of alluvial fans into types is made on the basis of
landscape observations. The following three types of fan were distinguished in the investigated
region.

Fans with Permanent Flow of Water over Their Whole Extent. The
surfaces of such fans are gently inclined plains, incised in the center by the stream channel.
The main area of the fan is at a considerable distance from the foot of the mountain, and slope-
wash processes have combined with alluviation of braiding streams to deposit the material. The

zone of recharge is high in the mountains. The inner part of the fan is composed of coarsely clastic material. Transition from the inner part to the outer apron, composed of fine material, is gradual in respect to decrease in slope and change in lithology. Groundwater, chiefly fresh, occurs at depths of 20-30 m at the outer margin of the fan and at 50-200 m in the inner part. Zones of water wedges are not clearly expressed in the landscape.

The plant cover of the inner part of the fan and the outer apron contains groups with dominant ephemeral and perennial members sustained by meteoric water. Over a large part of the fan, the main area of the outer apron, the surface is bare soil. The use of the plant cover for determining position of water table and mineralization of the water is limited in this situation.

Broad Fans of Perennial and Intermittent Streams, the Waters of Which Sink Completely into the Unconsolidated Deposits at the Head of the Fan. The source area of detritus for these fans lies in the high mountains. The fans extend for 10-20 km. The inner part is rather steep, 8-10°, and consists of rubble-gravel with a small amount of fine material. The surface of the outer apron is composed of sandy clay (loam) 1.5-2 m thick. The transition to the outer apron is sharply expressed in change of slope and lithology, extending across a width of 5-6 km. Groundwater in the inner part of the fan, the zone of groundwater in passage, lies at a depth of 80-100 m. At the upper edge of the outer apron the depth is 10-15 m, and at the lower edge 0.5-1 m.

Vegetation on the inner part of the fan, as with the type described above, is represented by groups characteristic of rubbly soil: the ephemerals wormwood, Russian thistle, and other saltworts. The plant cover on the outer apron consists of communities of phreatophytes nourished by groundwater. From the upper part of this apron downward, one may observe successive changes of communities corresponding to the depth to water and its mineralization.

Mud-Flow Fans. The surfaces of these fans are steep, slopes up to 20°, and are made up entirely of rubble-gravel. Outer aprons are practically absent. The plant cover over a great part of fans of this type consist of ephemerals with some feathery subshrubs.

The first two types of fans are promising for fresh water. The third type is practically free of water. Detailed study of the hydrogeological conditions of fans likely to contain water is made with a whole complex of methods.

In fans of the first type, zones with different hydrogeological conditions are distinguished on the basis of landscape features. The zone of greatest depth to the water table (zone of descending water) is distinguished by steep slope of the surface and by coarse-clastic material. The plant cover is thin, consisting of groups characteristic of rubbly deserts, the ephemeral wormwood. The transition to the zone with relatively deep groundwater (zone of discharge) is characterized by shallower slopes, smooth surface, and fine loamy material.

The main area of this zone is occupied by sown fields. Natural vegetation is found in small patches in the lower part of the zone, represented by communities of swamp species (sedge, rush, Bermuda grass) which mark sites of groundwater seeps. Because of full cultivation of man, landscape features that would permit fairly accurate determination of depth of the water table and mineralization of the water within each hydrogeological zone are absent over most of the area of this type of fan. These problems are solved by electrical-prospecting methods: resistivity and induced polarization.

The geoelectric profile is determined chiefly by depth of the water table and the type of rock. Mineralization of groundwater in fans of this type proves to be less than 1.0 g/liter according to hydrogeological sampling. Resistivity curves of type K (ρ_K) are characteristic of the inner part of the fan, composed of coarse-clastic material with the water table at depths down to 200 m. A high-resistance layer (up to 1000 ohm-m) corresponds to dry deposits of fine gravel. The drop in apparent resistivity on the right side of the curve is due to the occurrence of water-saturated rocks in the section. This type of curve is characterized by an appreciable slope of the descending branches of the ρ_K curves.

A quantitative interpretation of the ρ_K curves permits us to determine the depth of the water table and to determine roughly the purity of gravelly material by the resistivity value. For this it is assumed that the groundwater is fresh.

Dry fine and coarse gravels have resistivities of 600-2000 ohm-m, but for water-saturated deposits of this grain size resistivity declines to 200-500 ohm-m. The error of determining depth of water table is 10-15% on the average. No other distinct electrical boundaries are normally recognized on the resistivity curves.

The geoelectric profile undergoes considerable changes at the transition to the outer apron on the described type of fan. Although the principal type of resistivity curve remains the K type, type A curves are also encountered with shallow water table (down to 10 m). Values of apparent resistivity are considerably lower, generally not exceeding 100-200 ohm-m. For type K curves, the left low-resistance branch of the curve corresponds to surface loams with a resistivity of 20-30 ohm-m. Below these occur dry fine gravels with resistivities of 50-100 ohm-m. The occurrence of water in the section leads to a decrease in resistivity to 40-60 ohm-m. The error in determining depth of the water table does not exceed 10%.

With shallow groundwater, type A resistivity curves dominate. Increase in apparent resistivity with increase in supply centers is associated with the occurrence of fresh groundwater in the section of loam. As a rule, the purity of gravels increases with depth. The slope of the rising branch of the resistivity curve depends on the permeability of the aquifer. An increase in slope of the curve corresponds to an increase in permeability of the rock. An explanation of this is readily found when we consider that the zone of groundwater discharge on the fan generally corresponds to the zone of irrigated farming, and this leads to comparatively uni-

TABLE 1. Geoelectric Section of Unconsolidated Fan Deposits (Salinization Zone Absent)

Lithology	Resistivity value, ohm-m			Ratio $\dfrac{\rho_{\text{dry rock}}}{\rho_{\text{saturated rock}}}$
	Min.	Max.	Ave.	
Boulder—cobble deposits, dry	600	2000	800	1.7-3.0
The same, wet	200	500	400	
Cobble—granule deposits, dry	150	400	200	2.0-2.5
The same, wet	70	150	80	
Pebble—granule deposits, dry	50	150	70	1.1-1.3
The same, wet	40	120	60	
Fine pebble—granule deposits, dry	40	100	50	1.1-1.3
The same, wet	30	50	40	
Pebble—granule deposits with loamy matrix, dry	20	50	30	0.3-0.5
The same, wet	40	100	70	
Fine pebble—granule deposits with loamy matrix, dry	10	20	15	0.03-0.1
The same, wet	30	70	50	
Clayey sand, wet	30	50	40	—
Loessial loam	10	20	15	—
Clay	3	10	7	—

form geoelectric conditions in the upper part of the section, whereas the rate of change in resistivity values is determined chiefly by the resistivity of the water-saturated rocks, which, in turn, depends on the purity of the gravels.

However, reliable quantitative interpretation of type A resistivity curves is difficult because of the weak differentiation of the curves. In analyzing extensive material for parametric measurements and in using electric well logs, we may represent the geoelectric section of the fan deposits in the indicated type of fan by Table 1.

From Table 1 it is clear that the resistivity method may be used for coarse determination of the composition of unconsolidated sediments. The resolving power of the method at a particular depth of the water table depends on the lithology. It is greatest for well-washed coarse and fine gravels (K = 2-3).

A decrease in size of clastic material and an increase in the content of clayey matrix appreciably lower the possibilities of the method for determining position of the water table, commonly making it impossible.

To obtain supplementary information concerning the section, it is necessary to use the induced-polarization method. This method is based on the well-studied relations of polarization and rate of potential drop in ion-conducting rocks to the moisture content and type of rock (Rokityanskii, 1959; Kuz'mina and Ogil'vi, 1965).

In contrast to the resistivity method, the induced polarization method involves two parameters that characterize the properties of the rock section.

1. Value of apparent polarizability $\eta_K = (\Delta V_{ip}^t / \Delta V_{pa}) \cdot 100\%$, where ΔV_{ip}^t is the potential difference of induced polarization for a certain moment of decay, and V_{pa} is the potential difference at the moment the polarization current is passed into the earth. We have calculated the value of η_K for a decay time of 0.5 sec.

2. The second parameter in the induced-polarization method is the rate of decay of the potential. The clearest picture of change of the time characteristics with change of potential difference was obtained for the decay characteristic of the ΔV_{pa} curve as the ratio $\tau = (\Delta V_{ip}^{10\,sec} / \Delta V_{ip}^{0.5\,sec}) \cdot 100\%$.

Apparent-polarizability curves are characterized by low polarizability values at low potential differences, corresponding to the zone of aeration (0.1-0.3%) (Fig. 1). With increase in potential difference, the values of apparent polarizability increase sharply to 6-10%.

The value of the decay rate proves to depend on the grain size of the gravel, increasing sharply (τ falls from 50 to 2) for coarse-clastic material. A quantitative interpretation of the apparent-polarizability curves was made by means of V. A. Komarov's set of graticules. The accuracy of determining depth of water table from quantitative interpretation of resistivity and polarizability curves may be judged from the data in Table 2.

A comparison of qualitative and quantitative interpretations of the polarizability curves and the decay-rate curves with sections of boreholes and test pits has revealed some relations between the polarization properties of the rocks and both their lithology and their water content. The results are shown in Table 3.

The induced-polarization method may be used to determine depth of the water table and to gain information concerning the lithology of the water-bearing rocks.

In fans of the second type, the surface of the inner zone, being the zone where groundwater is at some depth, has a greater slope than that in fans of the first type, also composed of coarse clastics. Transition to the zone of groundwater discharge is characterized by a sharper change in slope and in lithology of the surface material.

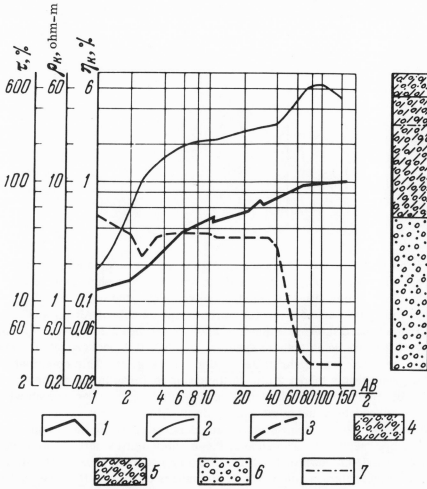

Fig. 1. Induced-polarization curves near boreholes. 1) Apparent resistivity curve (ρ_K); 2) apparent polarizability curve (η_K); 3) apparent decay rate of induced-polarization potential (τ); 4) loam with admixture of granules; 5) granule deposit with loamy matrix; 6) pebble gravel, flushed; 7) water table (1.5 m).

TABLE 2. Comparison of Accuracy in Determining Depth of Water Table by Resistivity and Induced-Polarization Data

No. of resistivity experiment	Depth of water table from well data (H)	Depth to water from resistivity curves (H_1)	Percentage error, $\frac{H - H_1}{H} \cdot 100\%$	Depth to water from polarization curves	Percentage error, $\frac{H - H_2}{H} \cdot 100\%$
1	1.5	1.5	0	1.5	0
2	1.0	0.5	50.0	0.9	10
3	7.1	6.5	9.0	7.4	5
4	14.2	14	1.3	13	8
5	13.5	6	—	13.5	0
6	1.8	1.2	30	1.9	6
7	11.0	14	27	11	0
			$\sigma_{ave} = 19\%$		$\sigma_{ave} = 4\%$

TABLE 3. Dependence of Polarization Properties of Rocks on Their Lithology and Water Content

Lithology	Polarizability, %, for decay time of 0.5 sec	Decay constant, $\tau = \dfrac{\Delta V_{ip}}{\Delta V_{ip}^{0.5\,sec}} \cdot 100\%$	$k = \dfrac{\eta_{aquifer}}{\eta_{supra-aquifer\ rocks}}$
Cobble-granule gravel		2-5	
dry	0.1-0.5		14-20
moist	0.5-2.0		
water-bearing	6-10		
Pebble-granule gravel			
dry	0.1-0.5	5-10	20-60
water-bearing	.6-10		
Fine pebble-granule gravel with loamy matrix			
dry	0.1-0.2	30-50	30-25
water-bearing	3-5		

The vegetation in the zone of groundwater discharge, as noted above, is clearly distinguished by appreciable density of the cover and the dominance of phreatophyte species. The distribution of the plant communities is zonal. An example is the fan of the Shaidan-Sai River (Tadzhik SSR). In the upper part of the outer apron occurs a belt of Alhagi and Peganum, indicating a belt of fresh groundwater at a depth of about 5 m. Downslope occurs a complex of Alhagi and Glycyrrhiza with no halophytic species. Against this background are found groups of Phragmites and the fresh-water shrub Halimodendron (Fig. 2). This combination of plants attests to shallow groundwater (about 2 m). Salinization features in the water are absent. Lower we find a complex of communities of Statice-Cynodon and "akbash"-Aeluropus with patches of Halimodendron and Glycyrrhiza and some Phragmites. This complex contains combinations of fresh-water phreatophytes (Glycyrrhiza and Halimodendron) grading into forms tolerating considerable salt (Statice, "akbash"). The zone of these communities outlines an area of slightly saline groundwater at a depth down to 1.5-2 m. Farther occurs a broad belt with a predominance of halophyte groups (Halocnemum) and communities of Alhagi-Halocnemum with patches of Halimodendron; i.e., we find the joint growth of individual relatively fresh-water species (Halimodendron and Alhagi) and species adapted to conditions of high salinity (Halocnemum). The groundwater here lies at a depth of about 1 m, and its mineralization, according to sampling data, is 3-4.5 g/liter.

In small isolated segments is found swampy-type vegetation, indicating direct seepage of groundwater at the surface.

The observed sequential change from fresh-water vegetation to halophytic is due to the gradual increase in mineralization of the groundwater toward the lower part of the discharge zone, a fact demonstrated by sampling.

The combination in the lower belt of strongly developed halophytes and relatively fresh-water species has been the basis for suggesting the presence of strong salinization of the upper soil layer with a relatively low mineralization of the groundwater. Such salinization of water and soil, in turn, attests to high transpiration of water by plants and evaporation from the surface of the ground with abundant inflow of fresh water, i.e., it indicates intense discharge of groundwater.

In the indicated type of fan, on the basis of landscape features, we may thus a) outline the zone of groundwater discharge by the association of phreatophytes, b) distinguish districts within this zone differing in depth of water table and mineralization of groundwater, c) predict the ratios of recharge to discharge of groundwater, and d) determine the amount of salinization of the upper soil layers.

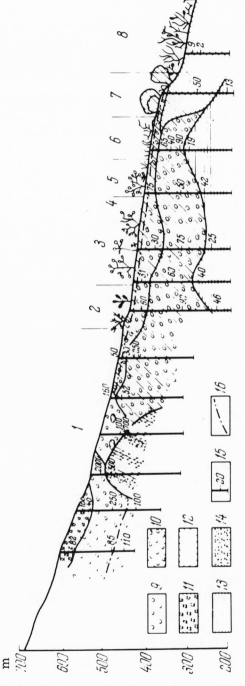

Fig. 2. Landscape-geoelectrical profile and cross section of the Shaidan Sai alluvial fan.
1) Ephemeral–Artemisia association, indicating a zone of deep groundwater; 2–8) associations indicating the zone of discharge; 2) Aeluropus–Alhagi association with some Glycyrrhiza and Halimodendron; 3) associations of Statice–Cynodon and "akbash"–Aeluropus with patches of Halimodendron, Glycyrrhiza, and Phragmites; 4) solonchak (salt bottom); 5) associations of Alhagi–Halocnemum and of Halocnemum with patches of Halimodendron; 6) association of Halocnemum with some Phragmites; 7) Alhagi–Halocnemum–Tamarix association; 8) Calligonum–Alhagi–Phragmites association on hills and Halocnemum–Phragmites in low areas; 9) well-washed granule gravels; 10) fine granules with loamy matrix; 11) clay with admixture of granules; 12) loam; 13) clay; 14) interbedded clay, sandstone, and limestone; 15) apparent resistivity, ohm–m; 16) water table.

The geoelectric features of the section of this type of fan have much in common with those of the first type of fan. In the headward part of the fan, where the depth to water table reaches 200-300 m, the resistivity curves are of K type. The geological-hydrogeological interpretations of the resistivity curves are similar to those discussed above.

The geoelectric profile of the outer apron reflects the shallow water table and the salt content of the zones of aeration and saturation.

Type A resistivity curves are but weakly differentiated. The apparent resistivity ranges from a few ohm-m at low potential differences to 100-150 ohm-m. To obtain information on the depth of the water table and on lithology, it is necessary to use the induced-polarization method.

A fundamental feature in using electrical-prospecting methods in the investigated example is the effect of salt content in the rock and water on the electrical resistance and the induced-polarization parameters. Discrimination of lithic boundaries in the section is possible only after preliminary isolation of zones having identical salinities of soil and water by means of the landscape method (see Fig. 2).

Fans of the third type are practically free of water, and are distinguished by a short profile and slight sorting of material throughout. The upper and lower parts consist of coarse clastics. The plant cover over the entire area of the fan consists of ephemerals. Almost all the investigated fans also bear feathery subshrubs, not phreatophytes. In the lower parts of the fans, the amount of such shrubbery increases somewhat and is better developed. As appears from two years of observation (differing sharply in amount of moisture), this fact is due to perched water in moister years. These fans are characterized by type KH resistivity curves with high values for all curves (500-1000 ohm-m). Minimum values of the curves generally indicate perched water or moistened rock. The ascending right-hand branch of the curve (slope generally of 45°) reflects the highly resistant base of bed rock.

On the basis of investigations indicated above, we may suggest a method of large-scale hydrogeological study of alluvial fans.

The first step in the study of the water supply on piedmont plains is analysis of the features of the region from map, literature, and unpublished data, including landscape and geophysical investigations if such exist. As a result of this stage of endeavor, preliminary landscape-indicator maps are prepared, and these should permit one to obtain a general view of the distribution of fans, to evaluate the source areas, in a general way, and the surface features, features with which the occurrence of water, vegetation, depth of water table, and mineralization of groundwater are associated.

Geophysical maps should permit one to determine the electrical properties of the rocks, to find characteristic values, and also to obtain information concerning the thickness of the unconsolidated material in the fans.

On the basis of data obtained in the first stage of the work, a grid for field traverses is laid out, and districts are set out for more detailed work. The second stage of investigation should be preliminary surveys of the traverses for landscape-indicator observations accompanied by resistivity surveys with electrode spacings no greater than 100 m, which will permit determination of the depth of the water table, mineralization of groundwater, and lithology of the upper deposits. As a result of this work, a map should be prepared to show types of fans according to landscape features. The third stage should be detailed multiple geophysical studies and, to a lesser extent, landscape-indicator studies in order to determine all hydrogeological characteristics for each type of fan distinguished.

LITERATURE CITED

Kuz'mina, É. N., and Ogil'vi, A. A., Possible Use of Induced-Polarization Methods for Studying Groundwater, Razved. Geofizika, No. 9 (1965).

Rokityanskii, I. I., "Laboratory study of induced polarization of ion-conducting rocks," Izv. Akad. Nauk SSSR, Seriya Geofiz., No. 7 (1959).

LANDSCAPE-INDICATOR INVESTIGATIONS OF KARST
A. G. Chikishev

The utilization of karst regions, in particular the construction of hydroelectric and industrial projects, the work of highway and civil engineering, raises complex problems for specialists in the various branches of science, the solutions of which are impossible without preliminary detailed study of the regions. Geographic-indicator investigations have become progressively more widely used in recent years, permitting more rapid, simple, and economic hydrogeological, engineering-geological, geomorphic, geochemical, and other studies, especially during the first reconnaissance stage.

Geographic-indicator investigations are based on a consideration of the systematic relations between individual physiognomic components of the natural complex or their combinations and decipient components of the landscape that are inaccessible to direct observation. In the various natural landscapes, the physiognomic and decipient components are different. But, most frequently, relief or vegetation, and their combinations, stand out as physiognomic components. In this respect there is interest in sinks, which, in forming irregularities in humid regions, affect the distribution of vegetation.

In geographic-indicator investigations the following basic trends may be noted: indicator geomorphology, indicator hydrography, indicator geobotany, and indicator landscape science. Indicator geomorphology uses relief as an index of the different characteristics of geologic structure of the region. In indicator hydrography, the drainage network stands out as an indicator of decipient components of the landscape. Indicator geobotany uses vegetation as an indicator* of the different physicogeographic conditions. Despite its newness, indication geobotany in respect to method has been more thoroughly developed than other trends, since geobotanists have "worked out in rather detailed fashion their views concerning indicators, their reliability, and the mechanism of indicator correlation" (Viktorov, 1966, p. 12).

Recently, greater significance has progressively attached to the problem of multiple use of all external physiognomic features of the landscape for determining decipient, chiefly morphological—structural features of a region. Indicator landscape science, resting on geomorphic, hydrological, geobotanical, soil, zoogeographical, and other data, permits us to examine more fully and all-inclusively, and to evaluate, the correlations among individual components of a natural complex.

---

* In indicator geobotany the concept of indicated object or feature is distinguished along with the concept of indicator. By indicated feature we mean a definite feature of a species or association having indication significance, although the species itself or the association is not the indicator (Viktorov and Vostokova, 1961).

During geographic-indicator and, in particular, karst investigations, special significance belongs to the problem of reliability of the indicators. By reliability we mean the degree of linkage between indicator and indicated object. Unfortunately the problem of evaluating the reliability of indicators and of the principle of their extrapolation has not yet received sufficient elucidation in the literature. Work on this problem is complicated by the rather insignificant amount of study devoted to the complex of interrelations between individual components of the landscape or to natural complexes of different ranks.

The question of indicator reliability has been examined most thoroughly by Viktorov and Vostokova (1961), who have proposed a five-division scale for evaluating this index (at 100% coincidence between indicator and indicated object, the degree of reliability is highest; greater than 90% it is high; between 75 and 90% it is satisfactory; from 60 to 75%, low; and below 60% it is negligible). The names denote that the reliability of indicators depends on the extent and character of the indicator, and also on the extent of the indicated object.

Landscape-indicator investigations of karst include questions on indication, by physiognomic components of the landscape, of the basic conditions of karst formation and on prediction of karst processes, and also on the appearance of various natural factors by the character and features of karst development. Karst-indicator studies are based on an investigation of the correlation between individual components of the natural complex, a correlation that is generally made either by comparing two key segments similar in their natural conditions (in one of which there is an indicated object) or by means of composite physicogeographic profiling (Chikishev, 1960a, 1960b).

In multiple landscape-indicator karst investigations, aerial photography and direct aerial observation are of fundamental importance. The use of air photos and preliminary aerial flights over the region make it possible to obtain the most complete information concerning the extent of karst development in the region, the morphological aspects of karst forms, and the hydrological conditions of karst formations without laborious surface work (Kiryushin, Bagrova, and Starichenkov, 1967). In mountainous regions karst in limestones that crop out at the surface is reliably recognized by a characteristic variegated-porous picture of the photo image and pitted microrelief, emphasized by the darker tone of vegetation associated with the low damper segments (Fig. 1).

Karst-indicator studies have appeared only recently. The first indication of the possibility of using one landscape component, vegetation, as an indicator of groundwater during hydrogeological investigations is found in the work of Savarenskii (1933).

Grebenshchikova (1939), in studying a swamp in the Ivanovo Oblast, distinguished ancient sinks by means of vegetation, sinks now filled with peaty deposits but having no expression in the present relief. Spore—pollen analysis has established an Atlantic age of the sinks, a much greater age than the surrounding swamps, since peat in the lowermost parts of the swamps began to form only in sub-Atlantic time.

Viktorov (1955a) established in Ust-Urt the systematic relationship of wormwood—shrub communities (Artemisia terrae abbae Krash., Atraphaxis spinosa L., Alhagi pseudoalhagi (M.V.) Desw.) with saucer-shaped sinks and large shrubs and trees (Tamarix, Elaeagnus angustifolia L.) with karst valleys. In another work Viktorov (1966) distinguished several stages in the development of surficial karst forms in Ust-Urt, clearly revealed against a background of stony desert by the continuous distribution of vegetation in them, which therefore appears to be a reliable indicator of karst regions.

Great interest is aroused by the work of Parfent'eva (1959), in which a method is discussed for indication of ancient sinks in Central Karatau by vegetation, sinks that are masked by recent deposits. Within ancient karst zones, with sinks filled with Cretaceous and Quater-

Fig. 1. Mountains in the southern part of the USSR. Karst in surface limestones, reliably indicated by the characteristic spotty-porous picture and
the pitted microrelief.

nary deposits, loose, well-drained, low carbonate soil has formed, covered with wormwood–
grass vegetation, differing sharply from the steppe variherbaceous–spirea and xerophytic
grass–variherbaceous vegetation of the neighboring regions, which do not exhibit karst topography. On Cretaceous bauxites occurs chiefly wormwood (Artemisia karatavica N. Krasch and
Abol.), and on the Quaternary deposits, containing more carbonate, are found associations of
silver-leafed wormwood (Artemisia juncea Kar. and Kir.), which permits us not only to recognize ancient sinks not expressed in the landscape, but to establish relative ages as well.

Sokhadze and Sokhadze (1961), in studying the vegetation of karst basins in the limestone
belt of Western Georgia, noted that many basins in the zone of beech–spruce–fir and spruce–

fir forests are treeless and that the vegetation of the basins has more of a high-mountain character than the surrounding zone. This is a reliable indicator of large sink basins on the southern slope of the Main Caucasus, and the basins may be readily outlined on air photos.

Zemtsov and Burov (1963) discussed important geomorphic-indicator features of promising gold placers in the upper reaches of the Tom River (Kuznetskii Alatau). They showed that karst processes, working under gold-bearing alluvium, do not favor the concentration of the valuable metal. For predicting the richest placers here it is necessary to know when the karst was formed, before or after deposition of the alluvium. Features indicating primary karst development of the bedrock are a smooth surface of the terrace and undisturbed occurrence of the alluvial layers, preserving fluviatile bedding.

An interesting attempt to use medium-scale air photos in karst-indicator investigations has been made by Kiryushin, Bagrova, and Starichenkov (1967), who prepared a table of relations between karst forms of the Pinega–Kuloi region and the relief and hydrogeological conditions and also a table of the interpretive features of karst forms and the peat bogs associated with them. These workers noted that deep steep-walled sinks are indications of a relatively thick zone of vertical circulation and deep occurrence of groundwater, whereas an abundance of lakes and sinkhole peat bogs attest to a comparatively high water table and to horizontal circulation of groundwater.

The above investigations lay a foundation, in great measure, for karst-indicator science, setting up the task of recognizing, by physiognomic components of the landscape, conditions of karst formation, predicting karst processes, and indicating various natural factors by peculiarities of karst development.

Indications of Karst-Forming Conditions and Karst Processes

Karst processes develop under a combination of definite conditions: presence of soluble rock, permeable rock, movement of water, and solvent capacity of the water (Sokolov, 1962). These conditions may be revealed in great measure by indicator methods of investigation, methods based on studies of the systematic relations between the various components of natural complexes.

Indications of Karst-Producing Rocks.
The occurrence of carbonate, sulfate, and halide deposits is an important matter in karst investigations, since the boundaries of possible development of karst processes are drawn primarily about areas where karst-producing rocks occur. Rock-indicator investigations acquire special significance in plains regions, where there is a comparative paucity of outcrops of such rocks because they are normally covered by unconsolidated deposits.

For indicating rock types, along with other landscape components, relief is widely used, since different types of relief are formed in different rocks because of response to erosion.

Under plains conditions, regions of carbonate rock are distinguished by comparatively weak dissection and by the development of chiefly deep canyon-like valleys characterized by rectilinear segments and by sharp elbow-like fractures in plain view. Within plateau areas of carbonate rocks, a rolling upland surface is characteristic, dissected by stream valleys, locally with stepped slopes (bench and slope) (Gogina, 1959). For mountainous regions of limestones and dolomites, sharply dissected steep-sloped relief is typical, with deeply incised streams following systems of tectonic fractions (Gvozdetskii, 1954).

An especially broad character of relief is used for revealing and mapping salt-dome structures. It has been established that salt domes in a locality correspond either to positive forms of relief or to negative forms near uplifts, appreciably dissected, consisting of a series

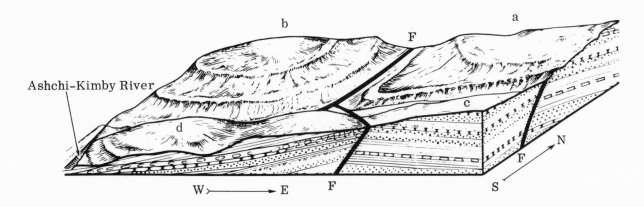

Fig. 2. Ural–Émba salt-dome region. Multiple dome, expressed in landscape by highly dissected features consisting of a series of uplands and lowlands (after Aristarkhova, 1956). a) Relatively uplifted northeastern limb; b) northern limb; c) southeastern limb; d) down-dropped southwestern block; F) fault line.

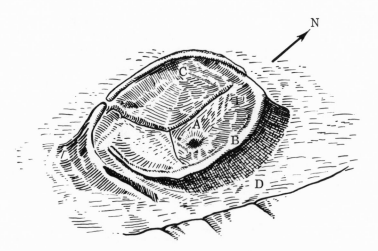

Fig. 3. Breached dome in the Ural–Émba salt-dome region (figure of M. I. Sokolov). A) Sink with exposed gypsum along walls, B) ring-shaped breached cuesta; C) inner slope of cuesta; D) plateau, consisting of clays.

of upland and lowland segments (Fig. 2). The uplands are formed chiefly on the limbs of the domes at places where more resistant rock outcrops. They generally have the form of cuestas, which are readily interpreted on air photos and represent one of the principal geomorphic features in mapping salt-dome structures (Aristarkhova, 1956). Breached domes are commonly manifested in the landscape by ring-like cuestas, which outline the central, relatively lower part of the dome, composed of sand–clay rocks (Fig. 3). In some places, basins, developed in the crestal zones of salt-dome structures, exhibit relatively well-developed surfaces (Korenevskii, 1961).

In making lithologic interpretations of air photos and using rock-indicators for specific localities, one of the quantitative features may be density of sinks, which, as investigations have shown, differ for different rocks. According to Golubeva (1953), the density of sinks in clays and anhydrite along the middle course of the Sylva River is many times (50-150) that in limestones and dolomites.

The boundaries of rock units may be indicated by surficial karst forms, which are commonly developed at the contacts of different formations. Thus, in the Kizel coal basin, chains of sinks may generally be traced along the contact between Viséan limestones and a coal-bearing sand-clay sequence, the latter serving as a confining impermeable layer for the permeable Viséan rocks (Pecherkin, 1963).

Along with rock complexes, it is possible to determine rock composition by morphological features. Long ago Kruber (1915) noted that typical sink holes at Krymskaya Yaila are developed only where pure limestones crop out, whereas in districts in which the limestones contain considerable admixtures of insoluble material, the sinks are strongly disrupted and covered with fine sediment. This indicator feature was later used by other investigators (Gvozdetskii, 1960, 1963).

In close connection with the type of relief determined to a great extent by dissection of the region, distinctive drainage patterns are developed, standing out as reliable rock indicators where rocks are poorly exposed. Carbonate rocks in mountainous regions are readily identified on air photos by the wide-spaced and broken drainage pattern. Crestal and near-crestal zones of salt-dome structures are commonly sites of denser stream patterns, a radial pattern, and concentric streams. An example of this concentric or encircling tendency is found in the swampy lake-like expanse along the lower course of the Zharly River (Ashche-Uil tributary) in the Ural–Émba region, due to local uplift associated with salt tectonics (Aristarkhova, 1956). With continuous uplift of salt masses and rise of the surrounding rocks, rising in the Caspian lowland at a rate of 2-3 mm per year, in connection with which the growth of many domes is almost unmarked by the effects of exogenetic factors (Kozhevnikov and Meshcheryakov, 1956), many streams are made to flow around the salt-dome structures. This is true of the Ural River below Ural'sk, where it makes a large bend around the Chalkar dome, and then bends sharply around the Inder salt dome. Similar large bends are found in the lower course of the Émba River, where this stream also bends around a salt-dome structure. The sharp swing of the Sula River at Romny is due to bypassing of the rising Romny salt dome.

For recognition of rock complexes, along with relief and hydrography, vegetation, reacting sensitively to the physical and chemical properties of the rocks, is also widely used (Chikishev and Viktorov, 1963). Gordyagin (1895), as early as the end of the past century, noted a substantial difference between the flora on limestone and that on granites exposed along the Ture River. The possible use of forest types for indicating rocks in the temperate belt was demonstrated by Vysotskii (1904), who established the restriction of spruce-fir forests to schists in the region of Kachkanar Mountain, of pine forests to olivine-bearing rocks, and mixed forests with significant amounts of Siberian pine (Pinus sibirica Mayr.) to gabbros and diorites. Later the selective role of plants relative to different rocks was noted by many investigators (Viktorov, 1955b, 1966).

In rock-indicator studies both individual species of plants and plant communities are used. The latter have higher indicator significance, since they may exist only under more restricted conditions. Individual species of plants are useful in indicating local objects. Plant communities are also a good interpretive feature, whereas individual plants, not generally visible on air photos, lack this quality.

The depth to which plants may indicate bedrock covered by unconsolidated deposits varies for different natural zones. As shown by plant geographers, the lithology may be indicated by plants in tundra regions to a depth of 1-2 m, in forest zones to a depth of 10 m, and in deserts to a depth of 20 m, since the deeper layers prove to have less effect on the vegetation. In relation to the principal types of karst formation, plants are divided into calciphytes (confined to carbonate rocks), gypsophytes (on gypseous rocks), and halophytes (on halide rocks).

The contacts between carbonate rocks and other rock varieties may be drawn with great reliability, because the plant cover on limestones and dolomites is distinguished by a clearly expressed physiognomic and floristic peculiarity. Basically, the stand of trees on carbonate rocks, because of the extreme dryness of the underlying rock and the harmful effect of calcium in excess, is normally thinner and suppressed than in neighboring districts underlain by other rocks (Gogina, 1959).

Of interest in this respect are the observations of Zagrebina (1964) in the Daldynskii region of the Yakutsk ASSR, where sedimentary (chiefly limestones and dolomites), volcanic (trap), and pyroclastic (kimberlite) rocks are found. The trap and sedimentary rocks are clearly differentiated by the character of the vegetation. On the carbonate rocks, regardless of the type of relief, occurs larch (Larix dahurica Turcz.), the principal forest tree of the region, in suppressed forms 8-12 m high, with thin, twisted crowns, whereas the larch reaches heights of 14-16 m on the trap and forms closed stands of timber. This difference is clearly seen on air photos, and this permits one to use it for interpreting type of rock. Another difference of the thin larch forest on and near carbonate rocks is the dominance in the soil cover of bushy lichens: Cladonia silvatica (L.) Hoffm., Cl. alpestris (L.) Raben., Cl. rangiferina (L.) Web., distinguished by light tones on air photos. In surface work, substantial aid may be derived from individual species of plants associated with some particular type of rock. For example, Cystopteris fragilis (L.) Bernh., C. Christens, Dryopteris robertiana (Hoffm.), and Woodsia glabella R. Br. are found only on carbonate rocks (Zagrebina, 1964).

On the western part of the Russian plain, on carbonate rocks covered by thin sand-clay deposits or residual material from the limestones, we find such plant communities as pine forests with oak and fescue (Festuca ovina L.) and pine—oak—beech variherbaceous (woodruff, asarum, oxalis, and other) forests. The greatest indicator value is held by communities of germander (Teucrium chamaedrys L.), verticillate sage (Salvia verticillate L.), and Italian aster (Aster amellus L.), which are indicators of the near occurrence of limestone. On the southwestern part of the Russian plain, on carbonate rocks, Preobrazhenskaya (1966) has found the following associations of oak and beech forests: Quercetum coryloso herbosum, Pineto-Quercetum coryloso-herbosum, Fagetum herbosum, Carpineto-Fagetum herbosum, Pineto-Fagetum herbosum, Carpineto-Fagetum sparsae-herbosum.

Carbonate rocks favor a shifting of steppe vegetation far to the north, and this is commonly a reliable indicator. In the Volga region, limestones are generally covered by pine forests in combination with sections of steppe. In the Northern and Middle Urals, numerous districts of steppe are spread widely in regions of spruce—fir and pine mountain-taiga forests where limestone is exposed, especially along stream valleys (Chikishev, 1964). Limestone crags in the mountainous parts of the Urals, according to Gorchakovskii (1960), are characterized by rocky, arctic—alpine and mountain—steppe species, among which many are endemic and relict: Shiverekia Kusnetzowii, Agropyron strigosum Boiss., Dianthus repens Willd., Dryas punctata Juz., Saxifraga caespitosa L., Dianthus acicularis Fisch, and Scorzonera glabra Kupr.

In the arid zone, outstanding indicators of limestone are Anabasis brachiata Fisch and Mey., A. truncata (Schrenk) Bge., Convolvulus fruticosus Pall., Salsola laricifolia (Turcz) Litw., S. chivensis M. Pop., and Nanophyton erinaceum (Pall) Bge.

The lower plants, as well as the higher, have important rock-indicator value. Among these plants, lichens, which exhibit close relationship with the substrate, are especially important. Merezhkovskii (1911) pointed out the restriction of the lichen Aspicilia fruticulosa Flag, A. affinis Mer., Acarospora coerulescens, and Lecanora azurea to carbonate rocks in the vicinity of Bol'shoe Bogdo Mountain in the Caspian lowland. On the Ust-Urt Plateau, communities with dominant Acarospora cervina Mass. and Aspicilia aspera (Mer.) Savicz. are widespread on exposures of Sarmatian limestones and their residual soils (Viktorov, 1960b).

Indication of Cretaceous rocks (chalk) by vegetation has been worked out less thoroughly, although such rocks are widespread, especially on the Russian plain and in Central Asia. Such questions have been worked on most thoroughly for arid regions, where Anabasis cretacea Pall., Lepidium Meyeri Claus, and Linaria cretacea Fisch are found on Senonian chalk in the early stages of soil formation, and groups of cretaceous semishrubs are found in the later stages (Vostokova, 1953).

Vegetation on gypseous rocks differs sharply from that on carbonate rocks, both in regard to species and in regard to ecological aspects, and this serves as a basis for indication of sulfate rocks that have been modified intensely by karst development. For example, the dominant forms on Upper Cretaceous and Paleogene gypsum in southern Fergana are Zygophyllum eurypterum Boiss., Reaumuria turkestanica Gorschk., Anabasis macroptera Moq., and Helianthemum soongoricum Schrenk. Here, Zygophyllum eurypterum Boiss, commonly indicates outcrops of Bukhara gypsum (Viktorov, 1949).

The possibility of indicating sulfate rocks in arid regions by lower plants has been demonstrated by Viktorov (1960a, 1960b), who established the restriction of gypsophilic lichens (Collema minor (Pasch) Tomin, Caloplaca bracteata Jatta) to unconsolidated weathering products of gypseous rocks, and also to low areas where gypsum has accumulated. He has also noted the massive development of Caloplaca paulsenii A. Z. in gypsum and gypseous rocks of Ust-Urt, where it forms a rather dense cover on outcrops of gypsum. Thus, in studying areas of gypsophilic lichens, we may outline gypseous rocks covered with loose soil and may also outline gypsum at the surface.

Halide rocks are clearly revealed by halophytic vegetation, capable of transmitting large quantities of readily soluble salts, principally chlorides. Sharply expressed salt anomalies associated with nearness to salt-dome structures are well confirmed by vegetation, which clearly indicates both degree and type of salt content in the substrate. In the northern Caspian region, for example, communities of halophytes are found on such salt domes (Suaeda confusa Sljin. and Halocnemum strobilaceum), whereas the surrounding plains are covered with meadows of salt-marsh grasses with Aeluropus litoralis (Gouan) Parl. and Tamarix ramosissima Ldb. Sometimes nearby salt-dome structures are exposed in districts with a background of plant communities of individual halophytic species (Shvyryaeva, 1964). Thus, the eastern part of the Azgir uplift in the northern Caspian region is marked by the appearance on its crest, in a wormwood community, such halophilic elements as Salsola laricina Pall., S. crassa M. B., Artemisia pauciflora Web., and Anabasis aphylla L.

It is possible to indicate by means of vegetation deep-seated salt domes if conditions obtain for migration of the salts to the surface. Such domes in the region of the southern Émba, for example, are reflected at the surface by well-defined districts of high salinization, in which occur saline soils and salt marshes with halophytic communities: Halocnemum strobilaceum M. B., Anabasis salsa (C.A.M.) Benth., and Atriplex cana C.A.M., readily marked against a general background of Agropyrum and white wormwood in surrounding districts with nonsaline or slightly salinized soils (Shvyryaeva, 1964).

Indication of Jointing in Rocks. Jointing in rocks, determining the permeability of the rocks, is one of the basic conditions for development of karst. During the Soviet regime, great success has been attained in the study and analysis of jointing in karst-producing rocks, and also in working up methods of investigating the joints of sedimentary rocks (Barkov, 1932; Gvozdetskii, 1954; Sokolov, 1962; Maksimovich, 1963; Charushin, 1963).

Four basic types of joints are generally distinguished: lithogenetic, tectonic, weathering, and mechanical unloading. During indicator investigations we shall most frequently have to do with tectonic and weathering joints.

Tectonic joints are characterized by wide distribution. They are commonly well indicated by vegetation and geomorphic elements. On Ust-Urt, for example, long lines of crowfoot overgrowth clearly mark subsurface fractures that have absolutely no expression in the relief (Viktorov, 1955a). Joints along chains of sinks, developed chiefly along tectonic joints, are also clearly marked.

Weathering joints are normally developed along tectonic joints in the upper exposed parts of karst-producing massifs, gradually working downward into the rock. They form an intricate network of fractures varying in density, width, and length. Weathering joints, the depth and width of which depend on the degree of weathering of the rock, are well indicated by vegetation, especially by bushes developed along the joints. Several physiognomically different states of weathering may be distinguished: moss–lichen, brush, and continuous plant cover. "Each of these stages corresponds to a thickness of the weathering zone, from a few centimeters for the first to several meters for the last" (Viktorov, 1966, p. 65). In carbonate rocks, weathering joints are sometimes found to depths of 50 m and more (Pecherkin, 1963).

Indications of Moving Karst Water. The problem of indication of karst water is very complex, and it has been but little studied yet. However, various landscape features are widely used during hydrogeological investigations that permit one to judge with satisfactory reliability the presence, depth, and mineralization of groundwater, and also to recognize some elements of the regimen and dynamics of this water.

The most important indices of definite hydrogeological conditions are relief (individual forms and types), vegetation, and morphostructural features of the region. Landscape analysis permits one to recognize not only higher aquifers, but also deep aquifers. Attempts have been made recently to use this method to identify confined water associated with definite morphostructural complexes and generally discharging in tectonically weakened zones.

Indications of the Aggressiveness of Karst Water. This important problem has not yet been given proper attention in either the domestic or foreign literature.

Indications of Karst Processes and Forms. The discovery of karst processes and forms by physiognomic components of the landscape, especially in regions where karst phenomena are not outwardly manifested, is an important task for karst-indicator investigations, since prediction of the development of karst and infiltration of groundwater is of fundamental significance for solution of a number of problems in the national economy.

It is possible to determine the intensity of karst processes and the degree of karst development by various features, chiefly of hydrogeological, geomorphological, hydrographic, and geobotanical character.

The development of concealed karst processes is indicated by springs and, especially, travertine deposits that accumulate at the mouths of underground streams. Travertine, or calcareous tufa, is widespread in the Caucasus, the Crimea, Central Asia, and in several other parts of the Soviet Union. Of interest, for example, are the numerous travertine deposits in the overthrust zone bordering the Kopet-Dag on the north. The formation of these deposits was associated with the leaching of Cretaceous limestones by meteoric water filtering down to great depth, encountering younger clayey rocks in the fault zone, acquiring upward movement, and flowing out at the surface. The wide distribution of travertine deposits along the line of a fault at the sites of active and already dried-up springs points to the intense development of karst processes in the northern Kopet-Dag, although karst phenomena have almost no external expression in the natural landscape.

Underground karst springs and associated caverns may be indicated by development of so-called thermal polinia (hole in ice) during the winter, when the temperature of the water in the spring is higher than the temperature of the water in the stream. In summer, on the other hand, the water temperature of the spring is generally lower than that of the stream.

There are now examples of using orographic and geobotanical data based on the study of plant successions, i.e., the change from one plant community to another, for determining the different stages of karst processes. Of interest in this regard are observations in the forest zone by Genkel' (1957) and in the arid zone by Viktorov (1966).

Genkel' (1957) used orographic features to define the stages of development of karst valleys in the Kungur district of the Perm Oblast. The first stage in development of sinks is characterized by scarcely noticeable settling of surface layers as a result of downwarping of beds over an underground cavern. At this stage, insignificant low places like poorly defined saucers may be detected by vegetation. This is of considerable significance since it permits us to predict the trend of karst processes and the development of sinks. The second stage is characterized by collapse of the roof into the cavern and the formation of a deep sink with steep slopes and irregular margins. Weathering and erosion gradually convert a collapse pit into a conical sink, on the floor of which one usually finds swallow holes, carrying water into the rock mass. At this stage the sinks are usually dry. The slopes are covered with sod or weeds (Urtica dioica L., Artemisia absinthium L., Carduus, Lappa). In the following stages of evolutionary development, in connection with clogging of the swallow holes and accumulation of detrital and peat material, the conical sink becomes shallower and acquires first a bowl-like and then a saucer-like form. Its gentle slope becomes covered with meadow grasses and herbs (Agropyrum repens (L.) P.B., Agrostis alba (L.), Poa pratensis L., Dactylis glamerata L., Bromus inermis Leyss., Sanguisorba officinalis L., Heracleum sibiricum L., Thalictrum minus L.). On the floors of bowl-shaped and saucer-shaped sinks a sedge biocenose usually develops (Carex gracilis Curt., C. vesicaria L., C. diandra Schrank, C. inflata Huds.). On the lower parts of slopes in the large bowl-shaped sinks, one sometimes finds bushes (Ulmus laevis Pall., Populus tremula L., Rosa acicularis Lindl). Saucer-shaped sinks, forming during continued accumulation by lithic material of older sinks, should be distinguished from such saucer-shaped depressions of primary origin, formed by settling of surface layers above underground caverns. With intense accumulation of peat, convex forms of microrelief may develop when the peat in the central parts of the sink basins rises above the surrounding surface. This phenomenon appears when the growth of the peat deposit is due to sphagnum moss. By means of pollen analysis, Genkel' (1957) established an Atlantic age for the investigated sinks and concluded that the processes of karst formation in the Kungur region were most active during Atlantic time, a time characterized by moist climate. At present most of the sinks are accumulating sand-clay material and becoming filled with peat.

According to Viktorov (1966), small karst depressions on the Ust-Urt Plateau are rather clearly distinguished by the continuous distribution of wormwood plants in them, the occurrence of which is associated with the greater moisture in this district. With development of saucer-shaped sinks, large bushes begin to appear in the center, and the stage of "sunken patch of wormwood" gives way to the stage of "bushy sink," and then to the stage of "composite sink" (Fig. 4). The presence of large "composite sinks" attests to extensive karst development in the region and to the development of underground caverns along with surface forms. Actually, in the center of some of these districts, Viktorov (1955a, 1966) noted sinks that formed as the result of collapse of a roof into an underground cavern that had developed in limestone beneath a sink. Establishment of the stage of karst forms supplies supplementary material for predicting the trend of further karst development.

Considerable interest is shown in the indication of ancient karst forms not expressed in the present relief. The possibility of indicating such forms by geobotanical methods has been demonstrated by Baranov and Ospoprivivatelev (1938), Grebenshchikova (1939), and Parfent'eva (1959).

Baranov and Ospoprivivatelev (1938), in their description of peat bogs in the region of Zelenodol'sk (40 km west of Kazan'), which occur in sinks and saucer-shaped depressions,

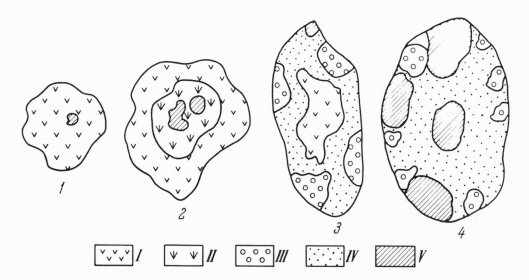

Fig. 4. Different stages of development of sinks on the Ust-Urt Plateau
(after Viktorov, 1966). 1) Stage of "sunken patch of wormwood"; 2) stage
of "brushy sink"; 3,4) composite sinks. I) wormwood; II) feather grass;
III) thick growth of atraphaxis; IV) thick growth of ephemeral grasses;
V) thick growth of prickly bindweed.

noted that one of them in an oak-linden forest is distinguished by a uniformly light green background of a continuous growth of sedges (Carex elongata L.) and various herbs (Camarum
palustre L., Lycopus europaeus L., Calamagrostis lanceolata Roth., Scutellaria galericulata L.,
Peucedanum palustre L., Ranunculus repens L., Naumburgia thyrsiflora Duby., Galium palustre
L.), clearly indicating saucer-shaped sinks of smooth oval form about 60 m across.

Grebenshchikova (1939), in studying a swamp on the divide between the Klyaz'ma and the
Nerl', where karst-producing formations of the upper Paleozoic are near the surface, had her
attention drawn to the different character of vegetation in separate parts of the swamp. In a
swamp forested by birch (Betula pubescens Ehrh.), round patches were observed, barren of
trees. In the herbaceous cover of these patches, the dominant forms were Carex rostrata Stokes
and C. stricta Good, and the dominant mosses were Sphagnum platyphyllum and Sph. Dusenii
C. Jensen, whereas the herbaceous cover of the swamp elsewhere was represented by the cotton
grass Eriophorum vaginatum L., Carex lasiocarpa Ehrh., C. caespitosa L., and Calamagrostis
lanceolata Roth., and the green moss Polytrichum strictum. Drilling revealed that these rounded
segments are sinks, filled chiefly with sphagnum peat.

Special investigations on the discovery of ancient sinks in the Central Karatau by vegetation were carried out by Parfent'eva (1959). It was shown that ancient sinks, covered by recent
sediments, may be readily detected and outlined by associations of wormwood (Karatau and
silver-leafed). Besides wormwood, these associations contain also small admixtures of Poa
bulbosa L., Festuca sulcata Hack., and Hulthemia persica Bornm.).

Karst as an Indicator of Natural Factors

Karst- and landscape-indicator investigations, along with discovery of the conditions of
karst formation, prediction of karst processes and forms, include problems of using karst as
an indicator of different natural factors. The close relations between karst and morphostructural
features of a region permit us to use the complex of karst features to indicate tectonic structures
of various orders and also the rate and trend of Late Quaternary and present movements. Attempts to use karst in prospecting for mineral deposits are of interest.

Indication of Tectonic Structures. Karst relief in great measure predetermines the morphostructural features of a region. This relation is seen most clearly in platform regions, where karst is best developed in the marginal zones of structural basins (synaclises) but is also present in anticlinal uplifts.

Geological, geomorphological, and special karst studies on Ust-Urt have furnished interesting material for indication of tectonic structures by karst forms. On Ust-Urt Plateau, on which Neogene carbonate and sulfate rocks occur, karst and karst-deflation forms, from small saucer depressions to gigantic basins, are widespread. The largest of these, in zones of regional downwarps, have clearly formed by overdeepening and expansion of tectonic depressions by exogenetic, chiefly karst processes (Geller, 1938; Luppov, 1948; Kopaevich, 1956; Kleiner, 1962). Basins of smaller size, as investigations of recent years have shown, are confined to the crestal zones of doubly plunging anticlines. Here, in connection with jointing in the limestones and better conditions for drainage of the water, karst processes developed extensively in Middle Pliocene time, and this led to destruction of Sarmatian limestones, armoring the surface, and to the formation of basins (Kleiner, 1962; Sakharov, 1967).

The morphology of basins associated with different tectonic structures differs. Basins formed on the crests of anticlines (Kaundy, Gurly, Asar, Karamandybas, Uzen', Shakhpakhty, Shordzha, and others) are distinguished by rather small size (up to 6-10 km in length), oval form, steep slopes, and sharp transition to the floor, whereas basins developed within local downwarps (Asake-Audan, Barsa-Kel'mes, and others) are usually not sharply outlined in plan, are not bordered their entire length by a scarp, and are characterized by indistinct transition from gentle slopes to the floor. The Karynzharyk basin, separating the Ust-Urt Plateau from the Southern Mangyshlak Plateau is of special interest. It occurs in a system of tectonic uplifts, and was formed by the joining of individual basins that formed on local anticlines (Kleiner, 1962).

Small sink forms – saucers, funnels, and "anas" (sinks having the form of a well, up to 30 m across and 50 m deep) – are found chiefly on the limbs of tectonic structures, where the rate of groundwater movement is greatest, furnishing the most favorable conditions for development of karst processes (Kuznetsov, 1965).

Thus, on Ust-Urt, as well as in other regions, an important factor in development and distribution of karst is the tectonic-structural aspects of the region and the hydrogeological conditions associated with these features. This permits us to use karst as an indicator of tectonic structures.

Indication of Quaternary and Recent Tectonic Movements. The development of recent karst and the redistribution of karst groundwater take place against a background of Quaternary tectonic movements, causing changes in the nature of the various natural processes and in the appearance of geographic landscapes. A study of Quaternary movements permits us to predict the intensity and trend of karst processes, and this has very important applications.

The possibility of indicating Quaternary tectonic movements by geomorphological geobotanical, and landscape features has been demonstrated by many investigations, the results of which are summarized in the works of Zvonkova (1959) and Viktorov (1966).

In conducting karst studies, ever greater significance is being acquired by methods that permit one to determine the introduction or removal of loose material clogging the sink cavity and to study the nature and rate of tectonic movements. Of interest in this respect is the work of Goncharenko (1956), who noted the widespread occurrence of empty cavities in Meotian and Pontian limestones in the region of the lower Dnieper, attesting to positive Quaternary movements in this region, favoring rejuvenation of karst forms and flushing them of loose material.

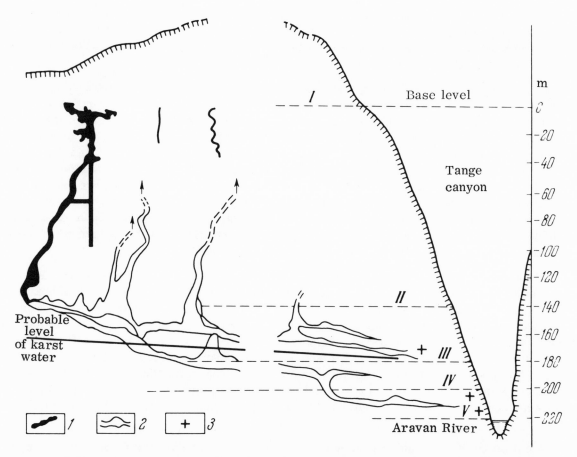

Fig. 5. Schematic east-west section through the Tyuya Muyun ridge and valley of the Aravan River (after Fersman, 1927). 1) Caverns, known; 2) caverns, inferred; 3) Kok-Bulak springs in Tange canyon; I-V) base levels.

Reliable indicators of crustal deformation due to Quaternary tectonic movements are horizontal systems of underground caverns, lying at different levels and correlative with stream terraces.

The method of recognizing Quaternary movements by study of fossil karst forms, which indicate uplift of segments of the earth's crust that have been subjected to karst processes, deserves attention. For example, the widespread occurrence of large sinks in Cretaceous rocks on the southern part of the Central Russian upland, filled with Paleogene and Neogene deposits, attests to intense development of karst processes at the boundary of the Late Cretaceous and the Paleogene, and also in the Neogene, when this region underwent general uplift, a fact marked by the formation of karst features at that time (Galitskii, 1963).

Indications of Mineral Deposits Associated with Karst. The effect of karst on the formation of ore deposits is dual. On one hand, karst forms are traps for heavier components; on the other, carbonate rocks are important as precipitants of metals under supergene conditions. Both factors may work in combination (Ginzburg, 1952). Nonmetallic deposits may also accumulate in sinks as well as metallic deposits.

Indications of iron, nickel, copper, lead, and zinc ores, or bauxite, phosphorite, diamond and gold placers, and other valuable deposits associated with karst represent a new and, in great measure, complex problem. For discovering such mineral deposits, methods of geographic-indicator and, especially, karst-indicator investigations are being more and more widely used.

A well-known example of investigation is the attempt to discover the richest deposits of rare metals at Tyuya Muyun, which occur in systems of ancient sinks along stream terraces. Fersman (1927), on the basis of the fact that stabilization of base level leads to prolonged karst development, concluded that the largest horizontal caverns might be associated with terraces of the Aravan River, at the corresponding levels. This led to an interpretation of the complex configuration of the underground caverns and gave precise prediction of the most promising sections (Fig. 5).

An interesting example of the use of geomorphic-indicator features for studying gold placers along the upper reaches of the Tom River (Kuznetskii Alatau) is described in the work of Zemtsov and Burov (1963). These workers established the fact that the richest placers are in karst depressions that formed before deposition of the gold-bearing alluvium. Secondary sink formation of the bedrock did not favor concentration of the metal. In this respect, it is necessary to discover the time of formation of the sinks, their age, in order to predict promising gold placers. Features indicating primary sink development are a generally smooth terrace, without collapse features or depression, and also undisturbed bedding in the alluvium exposed along the banks of the stream. On the other hand, the surfaces of terraces with secondary development of sinks exhibit large collapse features associated with younger processes, and the alluvium is strongly disturbed, locally displaying crumpling and rupturing of the beds in outcrops.

In searching for mineral deposits associated with karst, along with direct indicators, associated directly with the indicated object, indirect indicators are also used, the correlation of which with the indicated object is due to its relation to an element of the landscape that determines the distribution of the particular mineral deposits. For example, plant communities indicating carbonate rocks may become indirect indicators of a mineral deposit associated with limestones and dolomites.

Thus, individual investigations, even if still somewhat random, attest to the fact that karst phenomena and processes may be indicated objects and, under certain conditions, may themselves become indicators of some particular factors of the physicogeographic environment. Methods of landscape-indicator investigation of karst have not yet been worked out systematically, however. All actual indicators have not been isolated. Their reliability and value have not been determined, nor have regional indicator guidebooks been prepared, which might permit wide introduction of landscape-indicator studies of karst regions. All this is the task to be performed in future karst-indicator investigations, which are just beginning to be formed into a special independent discipline: karst-indicator science.

LITERATURE CITED

Aristarkhova, L. B., "Value of the geomorphological method for geologic mapping in the Ural-Émba salt-dome region," Uchenye Zap. MGU, Geomorfologiya, No. 182 (1956).

Baranov, V. I., and Ospoprivivatelev, N. Ya., "Geobotanical investigations of sinks and peat bogs in the Zelenodol'sk region," Uchenye Zap. Kazansk Gos. Zootekhn.-vet. Inst., Vol. 49, No. 1 (1938).

Barkov, A. S., "Karst of Samarskaya Luka," Zemlevedenie, Vol. 34, No. 1, 2, Moscow (1932).

Charushin, G. V., "Geologic method of studying jointing in karst-producing rocks," in: A Method of Studying Karst, No. 2, Perm (1963).

Chikishev, A. G., "Change of vegetation and soil of the Chusovaya River terraces in connection with the height of the terraces," Byull. MOIP, Otd. Geol., Vol. 35, No. 2 (1960a).

Chikishev, A. G., "Relation of vegetation to soil and hydrogeological conditions on the Chusovaya River terraces," in: Questions of Indicator Geobotany, Izd. MOIP, Moscow (1960b).

Chikishev, A. G., "Relation of the plant cover to climatic and soil-rock conditions in the Northern Urals," in: Plant Indicators of Soil, Rock, and Groundwater, Trudy MOIP, Vol. 8, Nauka, Moscow (1964).

Chikishev, A. G., and Viktorov, S. V., "Indicator geobotany," Priroda, No. 12 (1963).

Fersman, A. E., Morphology and Geochemistry of Tuya Muyun, Izd. AN SSSR, Leningrad (1927).

Galitskii, V. I., "Use of fossil karst forms for recognizing Quaternary tectonic movements," in: Method of Studying Karst, No. 4, Perm (1963).

Geller, S. Yu., The Origin of Interior-Drainage Basins, Problems of Physical Geography (Problemy Fiz. Geografii), No. 5, Izd. AN SSSR, Moscow-Leningrad (1938).

Genkel', A. A., "Peat bogs in sinks of the Kungur karst," Zemlevedeniya, Vol. 4 (44), Izd. MGU (1957).

Ginzburg, I. I., "Mesozoic karst and associated mineral deposits in the Urals," in: Weathering Crusts, No. 1, Izd. AN SSSR, Moscow (1952).

Gogina, N. I., "Analysis of the physicogeographic conditions for interpreting air photos during geologic mapping in eastern Siberia," Izv. Vysshikh Uchebnykh Zavedenii, Geologiya i Razvedka, No. 8 (1959).

Golubeva, L. V., "The density of sinks under different geomorphological conditions," Dokl. Akad. Nauk SSSR, Vol. 90, No. 1 (1953).

Goncharenko, M. G., "The lower Dnieper region," in: Karst in the Southern Part of European SSSR, Izd. AN SSSR (1956).

Gorchakovskii, P. L., "Preservation of plant relics and unique plant communities in the Urals," in: Preservation of Nature in the Urals, No. 1, Sverdlovsk (1960).

Gordyagin, A. Ya., "The vegetation of limestone crags on the Ture River in the Permian gulf," Trudy Obshch. Estestvoispyt pri Kazansk Univ., Vol. 23, No. 2 (1895).

Grebenshchikova, A. A., "Development of swamps in sinks of the Ivanovo Oblast," Sov. Botanika, No. 1 (1939).

Gvozdetskii, N. A., Karst, 2d ed., Geografgiz, Moscow (1954).

Gvozdetskii, N. A., "Investigations of karst regions," in: Methods of Geographic Investigations, Geografgiz, Moscow (1960).

Gvozdetskii, N. A., "Application of some methods during geomorphological investigations of karst," in: Method of Studying Karst, No. 2, Perm (1963).

Kiryushin, V. N., Bagrova, Z. A., and Starichenkov, I. P., The Image on Air Photos of Karst Forms of Relief and Associated Peat Bogs, Report of the Committee of Aerial Surveys and Photogrammetry (Dokl. Komiss. aéros"emki i fotogrammetrii), No. 3, Leningrad (1967).

Kleiner, Yu. M., "New data on the origin of interior-drainage basins," Dokl. Akad. Nauk SSSR, Vol. 147, No. 2 (1962).

Kozhevnikov, I. I., and Meshcheryakov, Yu. A., "Geomorphological methods of studying the subsurface," Priroda, No. 11 (1956).

Kopaevich, L. P., Tectonics and the Origin of the Sarykamysh Basin, Materials on Regional Geology (Materialy po Regional'noi Geologii, Trudy Vses. Aérogeol. Tresta, No. 2, Gosgeoltekhizdat (1956).

Korenevskii, S. M., "Salt karst of the Verkhnii Tis basin," in: Regional Karst Science, Izd. AN SSSR, Moscow (1961).

Kruber, A. A., The Karst Region of the Crimean Mountains, Moscow (1915).

Kuznetsov, Yu. Ya., "The karst of Ust-Urt," Zemlevedenie, Vol. 6 (46), Izd. MGU (1963).

Kuznetsov, Yu. Ya., "The Ust-Urt Plateau as an example of a karst region in the desert," in: Types of Karst in the USSR, Trudy MOIP, Vol. 15, Nauka, Moscow (1965).

Luppov, N. P., "Origin of the Sarykamysh basin," Izv. Vses. Geogr. Obshch., Vol. 80, No. 2 (1948).

Maksimovich, G. A., Principles of Karst Science, Vol. 1, Perm (1963).

Merezhkovskii, K., "A lichenological trip in the Kirgiz Steppe (Bogdo Mountain)," Trudy Obshch. Estestvoispyt. pri Kazansk. Univ., Vol. 43, No. 5 (1911).

Parfent'eva, N. S., "Vegetation as an indicator of ancient sinks of Central Karatau," Nauchnye Dokl. Vysshei Shkoly, Biol. Nauki, No. 3 (1959).

Pecherkin, I. A., "Study of karst development by drill cores," in: Method of Studying Karst, No. 2, Perm (1963).

Preobrazhenskaya, N. N., Aspects of the Distribution of Forest Phytocenoses in Connection with some Geological Conditions (from Data on the Investigation of Associations of Broad-Leaved and Pine Forests), Author's abstract of candidate's dissertation, Moscow (1966).

Savarenskii, F. P., Hydrogeology, ONTI NKTP SSSR, Moscow–Leningrad (1933).

Sakharov, S. I., "Basins of Mangyshlak," in: Problems of Physical, Economic, and Medicinal Geography of Kazakhstan, Alma-Ata (1967).

Shvyryaeva, A. M., "Possible application of the geobotanical method in prospecting for salt-dome structures in the northern Caspian region," in: Plant Indicators of Soil, Rock, and Groundwater, Trudy MOIP, Vol. 8, Nauka, Moscow (1964).

Sokolov, D. S., Basic Conditions for the Development of Karst, Gosgeoltekhizdat, Moscow (1962).

Sokhadze, E. V., and Sokhadze, M. E., "The effect of sinks on the plant cover of the limestone belt of western Georgia," Soobshch. AN GruzSSR, Vol. 26, No. 3 (1961).

Viktorov, S. V., "Types of gypseous deserts of Southern Turkestan," Byull. MOIP, Otd. Biol., Vol. 54, No. 1 (1949).

Viktorov, S. V., "Geobotanical features of karst-collapse processes in deserts," Byull. MOIP, Otd. Biol., Vol. 60, No. 1 (1955a).

Viktorov, S. V., Use of the Geobotanical Method in Geological and Hydrogeological Investigations, Izd. AN SSSR, Moscow (1955b).

Viktorov, S. V., "Geobotanical methods for geologic mapping and prospecting for mineral deposits," in: Methods of Geographic Investigations, Geografgiz, Moscow (1960a).

Viktorov, S. V., "Lichens of the Ust-Urt desert and their relations to some properties of soils and rocks," in: Questions of Indicator Geobotany, Izd. MOIP, Moscow (1960b).

Viktorov, S. V., Application of Geographic-Indicator Research to Engineering Geology, Nedra, Moscow (1966).

Viktorov, S. V., and Vostokova, E. A., Principles of Indicator Geobotany, Gosgeoltekhizdat, Moscow (1961).

Vostokova, E. A., Vegetation as an Index of Geological and Hydrogeological Conditions in Deserts and Semideserts in Connection with Their Utilization, Author's abstract of her candidate's dissertation, Moscow (1953).

Vostokova, E. A., "Use of geobotanical features in hydrogeological interpretation of air photos in arid regions of the SSSR," in: Questions of Indicator Geobotany, Izd. MOIP, Moscow (1960).

Vysotskii, N. K., "Some geobotanical observations in the Northern Urals," Pochvovedenie, Vol. 6, No. 2 (1904).

Zagrebina, N. L., "Reflection on air photos of the connection between vegetation and lithology in the Daldanskii region of the Yakutsk ASSR," in: Plant Indicators of Soil, Rock, and Groundwater, Trudy MOIP, Vol. 8, Nauka, Moscow (1964).

Zemtsov, A. A., and Burov, V. P., "A method of studying karst in prospecting for gold placers," in: A Method of Studying Karst, No. 2, Perm (1963).

Zvonkova, T. V., A Study of Relief for Practical Purposes, Geografgiz, Moscow (1959).

THE POSSIBLE USE OF LANDSCAPE INDICATORS FOR EVALUATING TRANSPORTATION THROUGH SWAMPS

L. A. Shevchenko

Landscape-indicator studies are of great significance for engineering-geological evaluation of a region, prospecting for groundwater, determining the suitability of land for agricultural utilization, evaluating the region for road construction, and so forth.

Of especially great significance, though still insufficiently evaluated, is landscape-indicator studies for determining the conditions of transportation through a locality. The necessity of studying conditions of movement through swampy terrain is accentuated by the utilization of new lands, including the Siberian swamps, where, because of the scant use of the region, it becomes necessary to transport long distances through swamps, with no roads.

The basis for landscape-indicator studies of swamps was laid by the works of Galkina (1955, 1959) and Ivanov (1957), who showed the relationship between the physiognomic aspect of a swamp and the conditions of its formation, its properties, the structure of the deposits, and the water-supply system. These views were later developed and refined by Romanova (1961) and Abramova (1964). Until now, however, their data have been insufficiently used for evaluating transportation conditions through swamps.

On the basis of interpreting landscape indicators by air photos, it is possible to evaluate the indicated conditions. Each swamp landscape, when it is examined on air photos as well as on the site, consists of several elements that may be single or may be repeated many times within the given landscape. These elements cannot be subdivided into component parts without changing to a different scale.

Such physiognomic elements, or swamp microlandscapes, constitute the physiognomic structure of the swamp landscapes. The simplest case will be that of a single physiognomic element or swamp microlandscape defining by itself the entire investigated segment. It is impossible to make any subdivisions here (for example, lowland reed, sedge, and other swamps in the forest-steppe and steppe zones).

If two or several groups of the simplest physiognomic elements (swamp microlandscapes) participate in the formation of the physiognomic aspect of the swamp landscape, we then have to do with a more complex structure or swamp mesolandscape. A swamp mesolandscape consists of the interrelations and interdependent combinations of swamp microlandscapes (lowland swamp, consisting of a combination of meadow grass, reed, and alder swamps).

When swamp mesolandscapes near each other join in one, a swamp macrolandscape is formed, or a more complex structural system (system of types of swampy tracts). Such systems may be observed in upland swamps where two convex mossy swamps combine or where transitional swamps are disposed along the gentle margins of upland swampy tracts (the northern part of the turf-podzol zone).

Each swamp microlandscape has its course of development and definite position in the relief of a swamp landscape of higher category, is characterized by its type of vegetation, has a definite type of surface, and is characterized by definite types of peat deposit.

The transportation conditions through swamp microlandscapes are made up of many factors. Besides the plant cover and the microrelief of the swamp, which represent one of the factors of transportation conditions, we must note a number of other properties: thickness and type of ground underlying the peat, the water table, the aqueous properties of the peat deposit, where the water naturally enters the peat, permeability, moisture content. The physicochemical and physicomechanical properties of the peat are also important. If we know the physiognomic elements of which the actual swamp landscapes we are studying are composed, i.e., if we know their physiognomic structure and for each of the microlandscapes have determined all or at least some of the above-mentioned features, then we may evaluate the transportation conditions through this area. In making this determination for an actual segment of any particular microlandscape, we may extrapolate our evaluation to all other segments of that same type of microlandscape. Thus, if in two swamp mesolandscapes of lowland type we observe microlandscapes with outwardly similar physiognomic structure (alder swamps with herbaceous cover), and in one of these we know all characteristics necessary for evaluating transportation conditions, we may speak then approximately of the transportation conditions in the other similar alder swamp. The similarity of physiognomic elements of swamp landscapes is usually well reflected by air photos, so that alder swamps having similar outward physiognomic structure in a locality will have similar direct interpretive features. The use of aerial methods opens up great possibilities for evaluating transportation conditions through swamps.

We should note that, in speaking of transportation through swamps, we have in mind not the construction of highways and not the setting out of courses with permanent tracks (or ruts) but rather of more or less episodic movement, perhaps but a single passage through the swamp, i.e., the case in which we are faced with primary utilization of swampy regions. The transportation conditions here depend chiefly on the properties of the upper layer of the peat deposit, and, in less degree, are determined by the deep structure of this deposit.

The first stage in determining the transportation conditions through swamps is the development of a rational classification of swamp microlandscapes. This classification must later be used as a landscape-indicator scheme for determining transportation conditions in different swamps.

In the field of establishing types of swamps, efforts have been expended chiefly in critical examination of existing classifications of swamps, partly by synthesis of such classifications and interpretation from the viewpoint of the objectives set up. In this work it is necessary to cover a large amount of material from the literature.

Standardization of swamp types must correspond to the following requirements: first, it must be sufficiently simple and easily applied; second, the discrimination of types must be physiognomically specific, to embrace all principal kinds of existing swamps, readily discernible on air photos, and at the same time to be distinguished by the complex of soil conditions; and third, it must contain the elements by means of which we may determine the type and species of swamp, i.e., the swamp landscape, by special maps (topographic, geobotanical, soil).

We discuss such a classification below in a somewhat generalized form. The proposed classification of swamps represents the summarized knowledge of many authors. In it are distinguished natural types of swamps, with reference to their source of water and mineral matter. Types of swamps are subdivided into species, and the principal species, encountered most frequently, are indicated. They may be readily distinguished from each other both by direct surface observation and by eye from the air, as well as by interpretation of air photos. For each species of swamp, we have pointed out the plant cover with its dominants, so that, by using the given classification, it is possible to determine the species and type of swamp by its plant cover. Each species and type of swamp occupies a definite position in the relief. Knowledge of this position for a swamp makes it possible for us to distinguish swamps as to types and even species by using topographic maps or air photos, even when it is impossible to use direct interpretive features. In the classification it is pointed out which type of swamp corresponds chiefly to any particular species of swamp soil. In the classification we have also noted the amount of moisture in the swamp in its natural state, which makes it possible to state what species of swamp of a given type is most difficult for transportation. The given classification does not pretend to complete systematic treatment of all existing data on swamps, but it may be used for particular applications within the framework of determining transportation conditions.

Generalized Classification of the Chief Swamp Microlandscapes for Transportation Conditions

I. Lowland (Eutrophic) Types of Swamps

Swamp Microlandscapes

1. Reed. Found in interior-drainage depressions of flood plains, along shores of water bodies, and on stream deltas. They form uniform light gray, almost white, sometimes dark gray tones on photos; they are structureless. Transportation is restricted, possibly only for equipment with high obstacle ability.

2. Sedge. Swamp with considerable participation of different species of sedge. Situated chiefly in stream valleys and in interior-drainage depressions on divides. Has cellular or honeycombed appearance (especially clear where land is permanently wet from outflow of groundwater). Where trees are present, we observe a uniformly fine-grained structure with an indistinct speckling of medium density. In other cases, no structure is observed, and the photos show a uniform, dark gray or gray tone. Transportation is basically possible; is restricted only for some types of vehicles. Movement through sedge—Hypnum swamps of this group is possible only for caterpillar-tracked vehicles and, in rare cases, for wheeled vehicles with extra-high obstacle ability.

3. Meadow grass. Swamps with participation of various grasses. Situated on flood plains of streams and lowlands about lakes. The photo image is inhomogeneous (ranging from light gray to dark gray), and it is structureless. Commonly, against the general background, one may note small individual light grains of indistinct form. Some images have a fine-grained warty appearance. Transportation is possible only for vehicles with caterpillar treads. In some of these swamps, wheeled vehicles with extra-high obstacle ability may get through (especially if the swamp is on a flood plain of a stream).

4. Wooded. a) Swamps with trees chiefly of alder. Situated on flood plains, stream terraces, slopes, and basins on divide areas. The structure is fine-spongy. Speckling is uniform, fine, very dense. The image is slightly indistinct, the tone uniform, gray or dark gray. Such swamps are relatively impassable. In the absence of herbaceous plants, transportation is possible only for vehicles with caterpillar treads and high obstacle ability. If a continuous cover of herbaceous plants is present, conditions for movement of vehicles are appreciably improved. Of wheeled vehicles, only those of extra-high obstacle ability might get through. b) Swamps with trees chiefly birch, occasionally with an admixture of pine. Situated on gently sloping deltas of flood plains, near terraces, on terrace slopes, and about the margins of most tracts of herbaceous swamps. The structure is uniform and fine grained. Speckling is dense, indistinct, slightly diffuse. The tone of the photos

is gray or dark gray. Swamps of this type are satisfactorily passable. Movement is free for vehicles with caterpillar treads, and in many swamps for wheeled vehicles with extra-high obstacle ability. With small amounts of moisture, transportation through the swamp becomes almost completely restricted.

II. Transitional (Mesotrophic) Types of Swamps

5. Mossy swamps with a predominance of sphagnum mosses (sedge–sphagnum, Hypnum–sphagnum). Situated on smooth divides, flood plains near terraces, gently sloping margins of upland swampy tracts, terrace slopes. Structure is either absent or is fine grained, homogeneous. Speckling is indistinct with diffuse margins. Design may be variegated. Tone of photo is irregular, light gray or gray. Transit possible chiefly for vehicles with caterpillar treads (possible limited transit of wheeled vehicles with extra-high obstacle ability).

6. Mossy swamps with pine and cotton grass (pine–cotton grass–sphagnum). Situated on smooth divides. Design fine grained, homogeneous. Speckling of average density, some slightly diffuse; grains of speckling rounded. Tone of photo gray or light gray. Transit possible for vehicles with caterpillar treads and some wheeled vehicles with extra-high obstacle ability.

III. Upland (Oligotrophic) Types of Swamps

7. Mossy swamps with trees, essentially pine (sphagnum with pine, brush–sphagnum). Situated on divides; occupy spaces near streams and canals. Structure fine grained, some medium grained or fine-spongy. Grains of speckling rounded. Design slightly smeared. Tone of photo light gray or gray, with large, dense stands of trees dark gray. Transit is possible for vehicles with caterpillar treads and some types of wheeled vehicle with extra-high obstacle ability.

8. Mossy swamps with no trees (open sphagnum swamps). Situated on divides. Structure absent; occasional small lines may be noted. Tone of photo light gray, almost white. Transit is possible chiefly for vehicles with caterpillar treads.

9. Swamps of ridge and low saturated ground (seep areas) or ridge and lake (sphagnum–cotton grass or sphagnum–scheuchzeria). Situated on divides. Have wavy-banded or mosaic design. Design may be concentric or concentric–wavy-banded. General tone of photo light gray or gray. Transit possible for vehicles with caterpillar treads, but only along the ridges. Transit through seep areas impossible for all vehicles, but limited transit possible through small zones of saturated ground, only for vehicles with caterpillar treads and high obstacle ability.

A special group of transitional varieties between present swamps and overgrown lakes or ponds forms spongy ground and floating mats. They are identified chiefly by sharp boundaries with open water, by dissected margins at the water boundary, by very light tones on the air photo with almost complete lack of structure, or by fine parallel streaking.

Some zones that do not fit the general classification are "top's," representing various elements of the drainage network of upland swamps (Romanova, 1961). These are characterized by elongate form, banded or reticulated design, diffuse indeterminate boundaries.

Spongy ground, floating mats, and "top's" are impassible for all types of vehicles.

Among the most important properties determining possible transit through swampy tracts, besides those listed above, is bearing capacity of peat, the permissible specific pressure per unit area of surface of the peat, expressed in kg/cm^2. The distribution of permissible specific pressure for different species of swamps is shown in Table 1.

TABLE 1. Permissible Specific Pressure (Bearing Capacity) on Different Species of Swamps

Lowland swamps	Transitional swamps	Upland swamps	Permissible specific pressure, kg/cm^2
Reed			0.25
Alder			
Sedge–Hypnum	Sedge–sphagnum	Sphagnum with pine	
Meadow grass	Pine–cotton grass–sphagnum	Sphagnum	
Alder–herbaceous	Hypnum–sphagnum	Sphagnum–cotton grass or sphagnum–scheuchzeria	0.25–0.5
Sedge	--	Marsh tea–sphagnum	0.5–0.75
Birch	--	—	
Sedge–tussock	--	—	
Drained swamps of all types	--	—	>0.75
Quaking bog, floating mat, "top'"	--	—	0.01–0.2

Note: Data taken with moisture in the range from natural content to saturation.

However, in investigations of swamps, by no means do we always find great tracts possessing uniform conditions for transportation. The complexity and patchiness of swampy landscapes are well known. The complexity of swampy tracts may be very high. Thus, if we examine upland swamps of the group having ridges with intervening low seep areas, we may discover primary physiognomic elements on the surface with very different characteristics for transit possibilities. The most favorable elements will be the ridges, and the most unfavorable, the extreme opposite, will be the zones of saturated ground (seep areas). In addition to these basic elements, there may also be secondary lakes, which are not only impassable but are also extremely dangerous, and there are also zones with artesian springs and tussocks of plant material near them, representing serious obstacles to transit. Along with negative features in regard to transit, we may also find positive features: mineral-based (inorganic matter) islands with fragments of forest, zones with abundant stumps, and so forth. The summed characteristics of a given swampy landscape, in the final analysis, depend on the relations of alternating ridges and seep areas, on the dimensions of these elements, the abundance of lakes and mineral-based islands, the size of these, and so forth.

On the slopes of convex (raised) peat bogs, great significance for movement of vehicles through the swamp will be the frequency of drainage lines. Such lines create serious obstacles to transit, even when the composition of the peat itself, from the viewpoint of transportation, may be more or less satisfactory.

One of the most complex elements of swampy tracts, from the viewpoint of transportation, is the lagg, or bog moat, the watery margin of the swamp. Laggs are most sharply expressed at convex (raised) upland swamps, where intense drainage of swamp waters to the margins of the swampy tract takes place. The boundaries of a lagg are clearly outlined on air photos, and also at the site, by the characteristic appearance of typical plant communities (groups of shrubs, water arum, cotton grass, and the like). As a result of studying swampy tracts, two rather large physiognomic elements of laggs have been distinguished: herbaceous and wooded. Herbaceous laggs are represented by sedge, horsetail, and reed varieties. They are composed of homogeneous, semiliquid, almost flowing masses of peat. In the wooded segments of a lagg, the peat is of inhomogeneous consistency. The lagg is not so moist under groups of trees, and it forms dense hillocks and ridges; in segments without trees, the lagg is like that of the herbaceous variety. Repeated alternations of lagg elements of both types may be found in a single swamp.

The inhomogeneity of the structure of swamps greatly complicates the evaluation of transportation conditions. The classification given above permits us to determine transit possibilities through a given tract as a whole, and, by following it, we may determine with good reliability the kind of transportation we should use in the initial stages of utilizing a given swampy region. In practice it frequently becomes necessary to determine transit possibilities only for a single route along which traffic will pass, but the task is frequently presented also of comparing several possible routes for choice of the most favorable for transportation.

The simplest, but by far the least accurate, method of approximate evaluation of traffic passability along a particular route is examination of the route on air photos. By this method, however, we may obtain only a general, very approximate view of the nature of the route itself. The complexity of accurate determination is magnified still more by the fact that relatively impassable elements on the surface of the swamp are generally diffuse and fragmentary: here we include districts with "top's" near drainage lines, seepage areas in areas of ridge and saturated ground, initial stages of formation of swamp creeks and secondary lakes. On air photos all such areas are represented by speckled or patchy zones or by discontinuous streaking. Therefore, to evaluate a marked route, we are faced not so much with the task of determining the position of one certain element that offers difficulty of passage, as recognizing areas of any particular density of such elements, i.e., determining the density of these elements. Only in this way can we work out the most rational variants for routes with avoidance of areas difficult to traverse.

The optimal method of precise determination of transit possibilities for an actual route is construction of special graphs along the route. These should probably be called Dobbs–Viktorov graphs, since they are based on frequency graphs used by Dobbs during geobotanical studies on Spitsbergen, and since Viktorov (1968) introduced certain changes into them, chiefly the replacement of frequency by density (number of specimens per unit area). For plotting the graphs, a strip of definite width is laid out on the air photo along the proposed route, and this is then divided into equal-area segments. A series of sample areas is thus obtained, adjoining one another. In these areas relatively impassable elements of the terrain are counted. We think it advisable to conduct this work with a subdivision of terrain elements into groups according to the ease they may be traversed by vehicles.

We have distinguished the following five groups of swampy microlandscapes, differing in trafficability (Table 2).

Group I. Impassable microlandscapes. This includes swamps the movement through which is impossible for all types of vehicles. It includes quaking bogs, floating mats, seepage zones (in swamps with ridges and seepage zones of saturated ground), secondary lakes, deep liquid "top's," and herbaceous laggs.

Group II. Swampy microlandscapes in which movement is poor, and those in which movement has restricted possibility for vehicles with high obstacle ability (chiefly vehicles with caterpillar tread). This group includes lowland reed and alder swamps, wooded laggs.

Group III. Microlandscapes with satisfactory (average) trafficability. Here we place swamps through which movement is possible for all vehicles with caterpillar treads and, in rare cases, for wheeled vehicles with extra-high obstacle ability. The group includes, of the lowland swamps, sedge–Hypnum, meadow grass, alder–herbaceous, birch–pine, and pine; and, of the transitional swamps, it includes sedge–sphagnum, pine–cotton grass, and sphagnum–scheuchzeria.

Group IV. Trafficable microlandscapes. Here we place swamp microlandscapes through which movement is possible for most types of vehicles, being restricted only for some

TABLE 2. Classification of Swamps According to Trafficability

Type of swamp	Swamp microlandscape	Trafficability group	
		with natural moisture content	moisture content at saturation and above
Lowland	Reed	II	II-I
	Sedge	IV	III-II
	Sedge–tussock	IV	III-II
	Sedge–Hypnum	III	II
	Meadow grass	III-IV	III-II
	Alder	II	I
	Alder–herbaceous	III	II-I
	Birch	III-IV	III-II
	Birch–pine	III	II
Transitional	Sedge–sphagnum	III	III
	Pine–cotton grass–sphagnum	III	II
	Hypnum–sphagnum	III	II
Upland	Sphagnum with pine	III	III
	Sphagnum	III	III
	Marsh tea–sphagnum	III-IV	III
	Sphagnum–cotton grass and sphagnum–scheuchzeria (ridge–saturated ground and ridge–lake complex)	III-II	III-II
	Quaking bog, floating mat, "top'"	I	I
	Drained swamps of all types	V	Depending on species of swamp

types with low obstacle ability. This group of swamps includes lowland sedge and sedge–tussock species, and, in some places, meadow grass and birch. Marsh tea–sphagnum upland swamps sometimes fit into this group.

Group V. Microlandscapes with good trafficability. This group includes drained swamps of all types. With deviations in moisture content of the peat deposits (from natural conditions), trafficability through the zone may change. In this case, swamps of a lower group may change to a higher, and vice versa. With increase in moisture content of peat, up to saturation and beyond, almost all species of swamps shift to a lower trafficability group. During drainage or drying out, they shift to a higher group.

Computations are made differently for each group: quaking bogs, floating mats, "tops's," seepage areas, lakes (impassable zones) are considered individually, and the remaining elements are calculated separately according to the characteristics given to them above. The computations are shown graphically, the abscissa representing distance from beginning of the profile to a given area, and the ordinate representing density of the elements in this area constituting difficult transit. As a result we obtain a series of curves, the maximums of which designate the sites of most difficult transit. Greatest attention here should be given to the curve showing the greatest concentration of impassable elements of the swamp.

We found it possible to test our views, in limited measure, concerning evaluation of trafficability at several swamps in Poles'e and the swamps of the Yaroslava region, and also to use data collected by Viktorov on swamps in the basin of the Pchevzha River at Volkhov. They confirmed our evaluations and permitted us to predict with creditable reliability the trafficability through swamps.

The wide use of aerial methods has made it possible to prepare trafficability maps for swampy tracts by means of the camera, with checking of only individual data. This has proved to be especially important for relatively impassable regions, where surface work is very difficult. It should be noted that color and spectrozonal photos give more reliable results for prediction than black and white prints.

As an example we may describe a swampy landscape for which we determined the trafficability by using both black-and-white and color air photos of medium and large scale.

The investigated swampy tract (the Kol'chikha swamp in the Moscow Oblast), occupying a broad flat basin open to the river, is a lowland swamp. Before field work was begun, preliminary interpretation was made by outlining the physiognomic elements and revealing their qualitative content. Interpretation was made on photos of medium scale (with refinements introduced from photos of larger scale). In drawing the boundaries of the physiognomic elements from air photos and then refining them on the site, almost no changes were necessary. In comparing descriptions of plant cover of physiognomic elements made from air photos with those made in the field, it was demonstrated that fundamental content of the elements was properly determined, except for some insignificant details that could not be distinguished on the air photos.

Seven physiognomic elements were distinguished in the Kol'chikha swampy tract during preliminary interpretation of air photos. Below we describe only some of these.

A. This element includes districts that appear on medium-scale air photos as structureless, dark gray in tone, with individual gray dots of various sizes. Lines corresponding to man-made drainage ditches are clearly recognized within the element. With close examination of the photos, it is possible to distinguish very small zigzag bands cutting across the element in different directions. It could be stated, as a preliminary opinion, that these features are either paths or natural drainage ditches. On large-scale photos the delineation of these features on the surface of the element is even clearer. In addition, the image of the photo displayed an inhomogeneous tone, not noted on the photos of smaller scale. The inhomogeneity of tone of the image may be explained by differences in moisture content of zones, expressed in changes of the small units (associations) of the plant cover. Inhomogeneity of color is especially well noted on color photos, which, for us, confirmed our opinion that inhomogeneity of the plant cover existed within this element. On large-scale photos it could be definitely stated that element A contains zones of greater and lesser moisture content, that the inhomogeneity of the plant cover is a measure chiefly of density and height of the herbage. Admixture of dry wood is possibly present in many places. Element A was interpreted as a district occupied by sedge associations with possible admixture of broad-leaved herbs, various swampy herbs, and grasses. Individual willow shrubs are noted throughout the district. On large-scale photos, areas covered by herbaceous vegetation also have distinct structure, consisting of indistinct lines, dots, and cells.

During surface studies of element A, it was found that the plant cover actually consists of different species of sedge. Moister sections (darker on the photos) are occupied by stunted sedges with a predominance of gray-blue sedge with a large admixture of various herbaceous plants. Zones with less moisture (lighter tone on photograph) are covered with plant species of sedge with admixture of spirea. The herbage is dense and high. Within the element are found pure growths of cattail. These may be identified in preliminary studies only on large-scale photos.

Thickness of the peat ranges from 35 to 75 cm. In the marginal parts of the element, at the contact with mineral-based islands, the thickness of the peat is 35 cm. The microrelief here is characterized by tussocks. Groundwater stands at the surface. The depth of standing water is 10 cm. Mineral soil is fine-grained, clayey sand with admixtures of granules. In the remaining part of the district, the thickness of peat is 70-75 cm. Mineral soil is clayey fine-grained sand with seams of clayey loam. The microrelief consists of sedge tussocks 10-15 cm high. Groundwater stands on the surface between tussocks. The average density of the peat (determined by a small impact tester of the Highway Scientific-Research Institute (DORNII)) is three strokes. Trafficability within element A is unfavorable (group II in our classification). Watery segments occupied by stunted sedge are most dangerous. At the contact with mineral-based islands, movement of vehicles is free.

B. In this element are included districts represented on medium-scale photos by inhomogeneous tone, but with a dominance of light tones (from light gray to whitish), and that are structureless. Individual light spots are clearly seen against the general background. The element is cut by darker straight lines (drainage ditches). White diffuse dots are scattered throughout the element, and these might be taken for haystacks on small-scale photos. On large-scale photos these white diffuse dots were interpreted to be man-made refuse areas.

By color air photos and by large-scale prints, one may see that the plant cover throughout the entire district is also inhomogeneous. Darker spots are distinguished, apparently moister areas than the surrounding district, and covered in part by other plant groups. We may definitely say, however, that element B has a completely different plant cover and much less moisture than element A. Zones with somewhat different vegetation may be pointed out most precisely on color photos. On large-scale photos, the distinctive structure of the district may be recognized (like very fine diffuse light speckling), and banding is clearly seen (banding is notable also on photos of smaller scale). In preliminary interpretation of element B, it was considered to be a district covered with associations of various herbs and crowfoot with possible admixtures of sedge and grass. Willow shrubs are common, chiefly along the ditches.

During surface studies it was shown that element B consists of meadows of various herbs, crowfoot, and lady's mantle. The plant cover in the district has been injured and is commonly secondary. Along old drainage ditches we may find stunted willow shrubs and growths of spirea. Within the element occur small moister patches, covered with sedge–hairgrass groups (they are also readily distinguished on large-scale air photos, especially on color photos). The whitish diffuse dots proved to be, as suggested, man-made dumps of light yellow sand (no haystacks or remnants were discovered). No microrelief was expressed in the entire area. Small tussocks were found only within sedge-hairgrass patches.

The thickness of peat throughout the district ranges from 1.2 to 1.5 m. The peat is brownish, highly decomposed, considerably desiccated, and this has led to a change in plant cover. Mineral soil consists of medium-grained clayey sand with seams of clayey loam. The average density of dry peat (with the impact tester) is eight strokes. The average density of peat in moister parts of the district is three strokes. A band of meadow, immediately adjacent to a wooded tract on a mineral-based island, cannot be differentiated on air photos from the remaining part of element B, but, in reality, it is not a swampy district but a moist meadow (peat horizon is absent). At the surface occurs a turf–humus layer 12 cm thick, and below that a layer of fine-grained clayey argillaceous sand.

Movement of vehicles through element B is perfectly possible. For vehicles with low obstacle ability, difficulties might be encountered only at drainage ditches and the wettest sedge–hairgrass patches.

The present paper by no means exhausts all questions of this important trend of investigation. But it permits us to state that landscape-indicator methods of evaluating trafficability where there are no roads, according to the difficulty of moving vehicles, represent a promising field for application of indicator investigation.

LITERATURE CITED

Abramova, T. G., "Indicator significance of the plant cover of swamps in the Leningrad Oblast,"
    in: Plant Indicators of Soil, Rock, and Groundwater, Trudy MOIP, Vol. 8, Nauka, Moscow
    (1964).

Galkina, E. A., Swampy Landscapes of the Forest Zone. Geographic Collection, Vol. 8. Ques-
    tions of Aerial Photographic Surveying, Izd. AN SSSR, Moscow (1955).

Galkina, E. A., "Swampy landscapes of Karelia and the principles of their classification," in:
    Peat Bogs of Karelia, Trudy Karel'sk Fil. AN SSSR, No. 15, Petrozavodsk (1959).

Ivanov, K. E., Principles of Swamp Hydrology in the Forest Zone, and Calculations of the Water
    Regimen of Swampy Tracts, Gidrometeoizdat, Leningrad (1957).

Romanov, E. A., Geobotanical Basis of Hydrological Study of Upland Swamps (by Using Air
    Photos), Gidrometeoizdat (1961).

Viktorov, S. V., "Landscape-indicator investigation and the morphometric study of landscape,"
    in: Aerial Surveying and Its Application, Transactions of the 9th All-Union Conference on
    Aerial Surveying (Trudy IX Vses. Soveshaniya po Aéros"emke), Moscow (1968).

THE SIGNIFICANCE OF LANDSCAPE EVOLUTION IN THE KARYN-YARYK BASIN FOR INDICATING PRESENT-DAY TECTONIC PROCESSES

L. F. Voronkova and T. I. Deryabina

One of the timely trends of indicator investigations is the use of this technique for evaluating the present-day evolution of landscape, particularly the forms that arise from the probable effects of present-day tectonic processes. Indication of natural processes is a complex problem, having been hardly more than stated thus far. For its successful solution it is necessary to work out techniques of dynamic indication of individual objects in order to accumulate data, the generalization of which may aid in solving the whole problem. Below we discuss some observations made in 1967 during geobotanical-indicator surveys in the Karyn-Yaryk basin (Western Kazakhstan).

Karyn-Yaryk, in the central part studied by us, from Kara-Maya Mountain on the north to Sumbya Hill on the south, is a depression bordered on the east by the Ust-Urt scarp and on the west by the bluffs of the Kinderli–Kayasansk Plateau. Immediately below the Ust-Urt scarps, in the easternmost part of the basin, occurs the huge solonchak (salt marsh) of Kinderli. West of this are the Karyn-Yaryk sands, separated from the Kinderli solonchak by a narrow belt of clayey plain, strongly dissected by gullies. Farther to the west lies a belt of complex desert (partly clayey, partly stony), rising gradually westward toward the Kinderli–Kayasansk Plateau. Through this belt are scattered numerous small clay-desert zones (takyrs) and isolated residual uplands (Ak-Mayak and other peaks). Thus, in going from east to west, the part of the Karyn-Yaryk studied by us may be divided into the following landscape microregions: 1) Kinderli solonchak or salt marsh, 2) clay plain bordering the solonchak, 3) sands, 4) clay deserts (takyrs), 5) complex desert.

The first information concerning the Karyn-Yaryk basin appeared in the seventies of the past century, when the Bis-Akta wells in the northern part were made the starting point for the movement of Russian Armies through the Ust-Urt to Khiva (see Grodekov, 1883). This information however is very fragmentary, and for our studies mention need be made only of the report of distribution of phytogenic tussocks (chukalaks) in the Bis-Akta region. In 1926 Karyn-Yaryk was crossed by a division of the Kazakh Expedition of the Academy of Sciences, SSSR, under the guidance of G. I. Dolenko (1930). The result of this survey was the first sketch of the Karyn-Yaryk landscape, containing information on the distribution of all the principal plant communities and soil varieties, on the appreciable induration of the sands, on the lack of water, and on the development of karst processes in the southern part of the basin. The Kinderli solonchak and the plain beyond the sands were described very sketchily.

Beginning in 1955 and up to the present, Karyn-Yaryk has been studied in detail in its geologic relations by parties of the All-Union Aerogeological Trust and the Joint Southern Geological Expedition of the Academy of Sciences, SSSR. As a result of these investigations it has been concluded (Sholokhov, 1964) that Karyn-Yaryk has had a very complex geologic history and has been modified by strong tectonic activity, expressed in changes from subsidence to uplift and the reverse. During the field work, studies were made also of some geochemical aspects of the landscape. In the work of Viktorov (1964), for example, it is shown that the Karyn-Yaryk sands are very rich in gypsum and belong to a special subtype of sandy deserts: gypsum-sand deserts.

The geobotanical-indicator survey made by us over a large part of Karyn-Yaryk, accompanied by interpretation of air photos and long landscape-ecology traverses extending for many kilometers, led us to conclude that all the investigated region is in a state of active evolution, expressed in sharp changes, locally having a catastrophic appearance, in the soil-plant cover. We shall describe these features briefly below.

On the shores of Kinderli salt lake we observed a rather broad expanse of thanatocenoses (i.e., dead plant communities) of Halocnemum strobilaceum M. B. The greatest amount of this dead genus was observed at the margin of the solonchak north-northwest of the nameless islands lying in the southern part of the solonchak. The dead Halocnemum communities consisted of separate large shrubs. Around many of these were depositional hillocks of "pseudosand," typical of this halophyte and giving rise to a landscape of hillocky solonchak. On the shores of internal-drainage basins, we frequently observed gradual dying out of the Halocnemum in connection with the fact that individual hillocks during their growth approached others, and joined, and from the hillocky solonchak a single layer of pseudosand formed, on which other less salt-tolerant plants may take root. The zones of dead Halocnemum we encountered had nothing in common with this process. The dead plants here appeared to have died suddenly, in phase with the typically expressed hillocky solonchak, as it were. All the shrubs and near-shrub hillocks were completely isolated, displaying no tendency to join, which points not to natural evolution but to some sudden process. This much may be concluded from the feature of thanatocenoses: the process was associated with desiccation, since the near-shrub hillocks (usually moist) were almost dry from the surface to the base, and the upper pseudosand was even subjected locally to deflation. The hillocks reached heights of 0.3–0.5 m and diameters of 5 m (locally even more). Their borders, because of wind action, were somewhat diffuse.

Desiccation of the Halocnemum has clearly been a rather recent process, since it has not yet caused any substantial changes in the salt complex, and the relations of Cl' and SO_4'' concentrations are rather typical for Halocnemum localities (at a depth of 0.10 cm, the values are 74.6% mg-eq for Cl' and 25.05% mg-eq for SO_4''; at 10–50 cm, the values are 64.4% mg-eq for Cl' and 35.2% mg-eq for SO_4''). However, judging from the total salt content, a slight leaching occurs here (5.8% of the salts at a depth of 0–10 cm, 12.8% at 10–50 cm). Zones where the Halocnemum has died out most extensively are recognized on air photos by very dark rectangular shapes.

It has been difficult to explain the dying out of Halocnemum by any local effects, especially since the phenomenon has been observed at the mouths of gullies, where, it would appear, water supply should have been most favorable. We are left with the suggestion that the region has been subjected to incipient uplift along the western margin of the Kinderli salt lake and that the Halocnemum thanatocenoses are associated with this.

The Karyn-Yaryka sands form hilly-ridges with large deflation basins; the sands are strongly fixed. The floors of the basins contain communities of Artemisia terrae albae Krasch. and disseminated groups of Halozylon aphyllum Minkw. Iljin. On the slopes occur Artemisia terrae albae Krasch. and Agropyrum sibiricum (Will.) Beauv. On ridge crests the association of various species of Calligonum with Salsola ruthenica Iljin, S. arbuscula Pall., and still others occurs.

Along the western edge, the sands are bordered by a high ridge, strongly broken and modified by wind from the crest down to the foot. At the base of this ridge is found a chain of wells. The northernmost of these is the Akkuduk well, and to the south occur the Tyué-singirdy, Saksor-kuyu, Kos-Ashchikuduk. The wells are at the contact of the marginal ridge with the clay deserts (takyrs) to the west. The formation of the marginal ridge was apparently related to the wells found here, near which flocks have been concentrated, breaking up the indurated sand. The wells are clearly fed by discharge from the clay deserts (takyrs), since, according to borehole data, the first aquifer below the surface descends to the east, and, consequently, groundwater must flow down from the takyr belt to the sands.

Construction of profiles from east to west, from the Kinderli salt lake through the sands to the takyr tract bordering the sands on the west, disclose several distinctive features in the structure of the sandy landscapes. One's eye is struck first by the dying out of large bushy sand-swelling buckwheat plants (species of Calligonum) at the crests of the ridges. Huge, dead, darkened specimens tower above all the ridges, creating a landscape of buckwheat thanatocenose. The shape of the shrubs is very extraordinary. Normally they are represented by sessile, spherical, or hemispherical forms, but in the present example they are treelike in outline, and appear to be raised, as it were, above the surface of the sand. In examining dead shrubs, it was found that the cause of the destruction of the buckwheat shrubs was removal of the sand from about its roots, by wind, and complete exposure of the root system to considerable depth. This phenomenon was also responsible for the change in shape of the shrubs, since the base, normally buried in the sand, proved to be exposed, and this made the shrub appear taller and better shaped than it normally is.

As an example we may cite a description made 10 km north-northeast of the Saksorka upland. Here, sands with large and medium-sized ridges appear in combination with broad basins. The height of the ridges above the floors of the basins range from 7 to 15 m, and the slopes are on the order of 15-30°. The basins are up to 200 m across. On the ridges are found dead buckwheat (buckwheat thanatocenose). In the lower parts occur Salsola ruthenica Iljin, Aristida pennata Trin., and Carex physodes M. V. The buckwheat reaches a considerable height, 1.5-1.8 m. The lower part of the trunk and the roots are exposed for 70-120 cm, so that specimens stand on a carcass of roots, so to speak. Sands along the crests of the ridge are strongly winnowed and form small barchans, crowning the ridge. Secondary deflation basins lie between the dunes.

Zones similar to the one described occupy the entire eastern part of the tract, essentially forming a single system of dead buckwheat along the ridge crests. Westward the dead specimens become less and less abundant, till at about 2-3 km east of the marginal western ridge described above the buckwheat thanatocenose is no longer found.

Another distinctive process that has developed in the Karyn-Yaryk sands is intense introduction of gypsum into the sands in the deflation basins. It is clear that the deflation basins recently contained gray wormwood and black crowfoot (Haloxylon aphyllum). Specimens of the latter are rather large (up to 2 m tall), woody, and, in outward appearance, must be assigned to the form typical of zones with shallow ground water (Vostokova, 1969). There is no doubt that in the recent past moisture conditions were more favorable than now. At present both the gray wormwood and the black crowfoot are dead. The floors of the basins are occupied by carpets of black lichen and black moss (Collema minor and Tortula desertorum). Locally patches of yellow lichen carpets are found (Caloplaca braeteata). Of the higher plants, only the ephemeral Eremopyrum distans Nevski is present. The abundance of these lichen coatings attest to gypsum enrichment in the sands. The greatest concentration of gypsum is marked by the yellow-lichen patches (Viktorov, 1960). The process of gypsum enrichment embraces the greater part of the sands. In the eastern part it has affected the entire basin. On the west it has advanced considerably farther than the dying out of the buckwheat, and has reached the edge of the ridge.

Thus, within the Karyn-Yaryk sands the following processes may be recognized: a) young, present-day wind action of the ridge crests, leading to buckwheat comblike crowns; b) desiccated sands, causing destruction of black crowfoot; c) gypsum enrichment of sands, causing death of the gray wormwood and formation of the lichen coating. The causes of these processes may be explained variously. The suggestion that man was the cause is clearly excluded. If cattle were raised extensively in the sand region, the sands would have been moved by the wind first in the basins, where the herds would be concentrated, not on the crests of the ridges. Furthermore, the raising of cattle would have prevented the development of lichen coatings, which because of the slow growth of the lichens cannot tolerate the trampling of cattle. We can also refute the idea that the accumulation of gypsum in the sands depends on some type of migration of the gypsum from the nearby gypseous strata of the Ust-Urt, as was stated by Viktorov (1964). The introduction of gypsum from groundwater is excluded because the aquifer is dipping to the east, and the movement of water is consequently toward the east. Aqueous migration of salts in the opposite direction is hardly likely, therefore. Just as unlikely is the eolian sprinkling of gypsum in the sands, since, in this case, the gypsum should have accumulated not in the basins but on the crests and slopes of the ridges, shielded by the wind currents, i.e., chiefly in zones with eastern exposure. But nothing like this is observed. The total content of salts on the surface of the sands on ridge slopes amounts to 0.123% (according to analysis of samples taken to the north-northwest of the Saksorka upland), but at the foot of the ridge, where the ridge grades into the basin, the content is 0.71%.

That there has been very recent transformation of the landscape here deserves attention. Dolenko (1930), in speaking of Karyn-Yaryk, only briefly mentioned incipient wind erosion of the ridges in places, but did not note the comblike crowns of buckwheat or of crowfoot. Nor did he observe the universal development of lichen crusts or their displacement of wormwood tracts. It is obvious that all these phenomena have appeared in the past three or four decades.

The most likely explanation of what we have described above is that there is some common cause, something that may have activated wind action in the upper structural stages of the landscape and, at the same time, worsened moisture conditions, circulation of water, and migration of salts in the lower stage of the landscape. This cause may have been recent uplift. Such an assumption allows us to explain the active removal of sand on ridge crests, which in turn determines the exposure and desiccation of the roots of sand-favoring shrubs. And, at the same time, discharge from the takyrs to the sand region must have been weakened, worsening the conditions for growth of phreatophytes and, especially, species with high transpiration rates, such as black crowfoot. Lastly, the same cause may explain the retarded migration of salt. The more mobile chloride would continue to leach out of the sand, but the less soluble gypsum would become relatively concentrated in great degree, causing the death of species not tolerant of high gypsum concentrations (Zakrzhevskii, 1934).

A certain probability that this explanation is true is lent by the circumstance that east of the sands on the shores of Kinderly salt lake there exist features we have already described as depending on such uplift. When we compare all that we have written above, we may conclude that the evolution of landscape in the Karyn-Yaryk basin (to what extent may be judged by the described ecological series) furnishes some indication that the western border of the Kinderli salt lake and the expanse of sand lying to the east of it have been affected by rather active uplift in very recent time. This is still a hypothesis, however, based on landscape-indicator investigations, and it needs verification by ordinary methods of geological investigation.

LITERATURE CITED

Dolenko, G. I., A Brief Description of the Landscape Regions of the Western Ust-Urt and the Mangyshlak Plain, Data of the Committee of Expeditionary Investigations of the Academy of Sciences, SSSR, No. 26, Kazakhstan Series (Materialy Komissii Éksped. Issledovanii

AN SSSR, No. 26, Seriya Kazakhstanskaya), Account of the Work of the Soil-Botany Division of the Kazakhstan Expedition (Otchet o Rabotakh Pochv.-bot. Otryada Kazakhstansk. Éksp.) AN SSSR, No. 4, pt. 2, Leningrad (1930).

Grodekov, N. M., The Khiva Campaign of 1873. Action of the Caucasian Division, St. Petersburg (1883).

Sholokhov, V. V., "The origin and tectonics of the Karyn-Yaryk basin," Izv. Vysshikh Uchebnykh Zavedenii, Geologiya i Razvedka, No. 10, A (1964).

Viktorov, S. V., "Lichens of the Ust-Urt desert and their connection with some properties of soils and rocks," in: Questions of Indicator Geobotany, Izd. MOIP, Moscow (1960).

Viktorov, S. V., "Vegetation as an indicator of gypsum accumulation in sands of the Karyn-Yaryk basin," Izv. Akad. Nauk SSSR, Seriya Geogr., Vol. 4 (1964).

Vostokova, E. A., Geobotanical Methods of Prospecting for Groundwater in Arid Regions of the Soviet Union, Gosgeoltekhizdat, Moscow (1961).

Zakrzhevskii, B. S., "The effect of the gypsum process on development of root systems of desert xerophytes and halophytes," in: Agricultural Utilization of the Desert of Central Asia and Kazakhstan, Tashkent (1934).

PHYSIOGNOMIC LANDSCAPE FEATURES AS INDICATORS OF ORIGIN AND DEVELOPMENT OF THE LANDSCAPE
A. I. Spiridonov

In the complex of various methods utilized in modern geomorphology, one of the most important involves perception of the landscape by means of studying its physiognomic features. This method is based on the view that the internal structure of the relief, especially its origin and development, finds outward expression in very definite forms on the earth's surface. Consequently, the surface, being readily accessible to direct observation, may be considered to bear physiognomic indicators or diagnostic features, and the origin, development, and present-day dynamics of the relief are decipient features, or indicated objects. The last also include relief-forming factors, particularly geologic structure, which we frequently evaluate from its morphological expression.

The outward (physiognomic) aspect or geometry of the relief (Efremov, 1949; Devdariani, 1963, 1967) is characterized by a complex of morphographic and morphometric features. In recent years the number of works in our country on the geometry, chiefly the morphometry of relief, has grown, especially in connection with solutions of problems in structural and engineering geomorphology and with prospecting for oil and gas structures (Philosofov, 1960; Piotrovskii, 1965; Devdariani, 1966). However, geomorphologists still allot too little attention to the analysis and synthesis of purely physiognomic features of the landscape. Until now, for example, we have had no clear picture of the use of many terms and definitions concerning the outward aspects of the earth's surface, and we have had no systematic organization of morphographic and morphometric principles, a lack that makes it difficult to prepare and read geomorphological descriptions and maps. Study of the origin of landscape should therefore be supplemented by investigation of its outward aspect, which, at some stages, may acquire independent significance.

As is well known, the physiognomic features of the landscape are widely used for geomorphological and geological interpretation of air photos. Geologic–geomorphic interpretation of detailed topographic maps are made by the same features. Even with direct observation in the field, the outward aspect of the relief represents a very important basis for drawing conclusions concerning the geologic and geomorphic structure and development of the terrain. Such conclusions are only preliminary, of course, and acquire great reliability only when many other facts are also used, especially geologic facts. If it is impossible to obtain such supplementary information, the physiognomic aspect of the relief is the only, or almost only, source of information concerning internal structure. This is just where the matter of geologic and geomorphic study of the Pacific Ocean floor stands when the investigator, on a base of bathometric maps and profiling of the ocean floor, arranges supplementary data from geophysical investigations and bottom samples. Conclusions concerning the geology and geomorphology of the moon and some other planets are still based to a considerable degree on the outward aspect of the landscape.

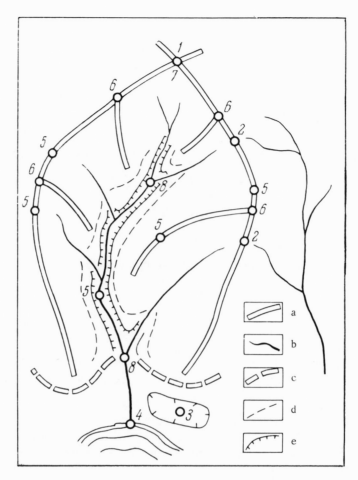

Fig. 1. Framework of landscape. Points: 1) apical;
2) saddle; 3) bottom; 4) mouth; 5) pivotal; 6) furcating; 7) intersecting; 8) confluent. Lines: a) divide; b) thalweg; c) hill-base; d) back inflection;
e) brow or terrace edge.

The relief of the earth's surface may be considered as a field of elevations, the smooth
representation of which gives a map in horizontal planes (Devdariani, 1966). This field is so
random, however, that analysis of the relief in all its complex variety may be done only tenta-
tively by breaking it down into a series of smaller component parts for which we may select
simple physiognomic characteristics and an approximate geometric shape. Such component
parts include (in order of complexity): elements of relief, simple and complex relief forms,
systematic groups (complexes or associations) of relief forms (Spiridonov, 1959). During geomorphological field studies, especially with observations from the air, with interpretation of
air photos and topographic maps, relief at first glance is perceived as a combination of forms.
However, subsequent analysis leads the investigator to smaller morphological units. The view-
er's concept of the outward aspect of relief is here augmented by morphographic and morphometric indicators, permitting him to expand and refine appreciably its characteristic, and this
enables him to formulate a generalized, synthetic picture of the relief of the investigated region as a whole.

The elements of relief are divided into a) points and lines, or the "framework of relief,"
and b) surfaces, or the "faces of relief" (Efremov, 1949).

Points may be considered according to their position in the profile and in plan. They are correspondingly subdivided into a) apical, bottom, mouth, saddle, edge, hill-base, and b) intersecting, confluent, pivotal, furcating (Fig. 1). Besides the position and forms of relief, each point is characterized by horizontal coordinates and absolute and relative elevation.

Lines may be a divide, crest, thalweg, hill-base, inner or outer edges. Each line is distinguished by a definite design in the profile and in plan. In profile, lines may be straight, convex, concave, concavoconvex, complex (wavy, broken); and bent lines may be arcuate, parabolic, elliptical, and so forth.

According to shape in plan, the following lines are distinguished: rectilinear, curvilinear, complex, and highly variable (annular, arcuate, parabolic, elliptical, U-shaped, serpentine, spire-shaped, pronged, and so forth). Along with outlines, the linear elements of relief are characterized by a series of quantitative indices. These include absolute and relative height, slope, azimuth of trend (general, or individual segments), radius of curvature, tortuosity.

The framework of relief as a whole depends on the mutual arrangement of the point and linear elements in vertical and horizontal planes: their absolute and relative height, density of distribution of the elements (number or extent per unit area, or their distance from each other), general design in plan view. These data are widely used for diagnostic purposes, since the framework of relief points to common petrographic-structural conditions for the formation of the earth's surface and permits us to evaluate the intensity of horizontal and vertical dissection of the landscape, the magnitude of recent movements of the crust, the ancient and recent structural plan of the region, the location and trend of folds, faults, and crushed zones, the jointing of rocks, and the alternation of weaker zones and rocks with more or less resistant zones (Gerenchuk, 1960; Shchukin, 1960; Nikolaev, 1962; Khain, 1964).

Valley lines, or thalwegs, are of special significance in indicator relations: their pattern in plane, their forms and development in longitudinal profile, their general and local gradients, and so forth. The drainage pattern (Fig. 2) is used as an indicator feature for geological interpretation of air photos, particularly for interpreting geologic structure (Petrusevich, 1962; Miller and Miller, 1961). The longitudinal profile of streams, pointing to the general conditions

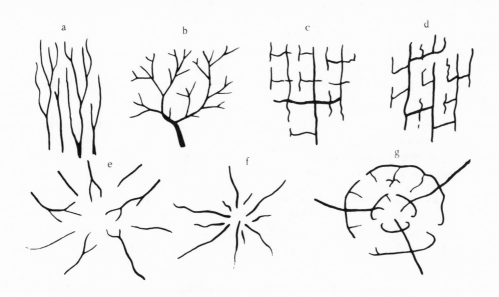

Fig. 2. Basic drainage patterns. a) Parallel; b) dendritic; c) rectangular; d) diagonal; e) radial; f) centripetal; g) annular.

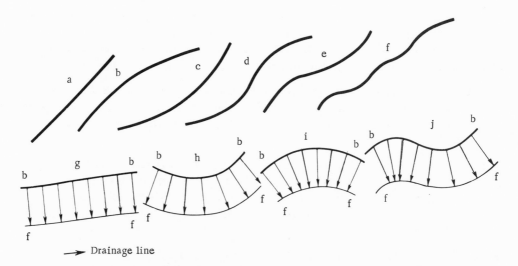

Fig. 3. Slopes, seen in profile and in plan. Slopes in profile: a) straight;
b) convex; c) concave; d) concave-convex; e) convex-concave; f) stepped.
Slopes in plan: g) rectilinear; h) convex; i) concave; j) tortuous. b-b)
brow of slope; f-f) foot of slope.

and stage of erosional development (Makkaveev, 1955) in places of anomalously steep or gentle
gradients, marks local tectonic activity of recent age, which appears on structure contour maps
drawn on the basement or deformed strata (Filosofov, 1960; Gvin, 1963; Volkov, 1964).

Faces of relief are divided into horizontal and inclined (slopes). Horizontal surfaces are
usually distinguished by weak general and local slopes, are apical surfaces of positive forms
(plakors), areas of terraces, or bottom surfaces of negative forms. In general form these may
be flat, convex, concave, or more complex (plane-wavy). They are characterized by the secondary aspects of absolute and relative elevation and by general and local slopes.

Horizontal surfaces are interpreted as elements of denudational beveling or of accumulation, or they may be stripped surfaces on a resistant stratum. Plakors may be considered
as relics of older denudational surfaces or of alluvial plains. Lower levels are associated with
succeeding stages of local erosional leveling. The distribution and number of stages of horizontal and subhorizontal surfaces are shown by plotting present-day profiles, distribution curves
of areas according to elevation (Tskhovrebashvili, 1964), or distribution curves of elevations.
Bimodal or, in general, polymodal distribution of elevations characterizes stepwise relief and,
in particular, the presence of planation surfaces.

Slopes, according to their forms in profile, are divided into straight, convex, concave,
concavoconvex, convexoconcave, and more complex (wavy, stepwise); according to their
shapes in plan, they may be rectilinear, convex, concave, tortuous, corresponding to parallel,
divergent, convergent, and mixed trends of drainage lines (Fig. 3).

The dimensional indicators, and also the slope shapes, include elevation, length, slope,
azimuth of trend line (strike and dip of beds), geometric pattern of curvilinear slopes in profile and in plan (arcuate, parabolic, elliptical, etc.), the radius of curvature of these, the ranges
and characteristic (modal) values of slope, strike, dip, and coefficients of tortuosity of the
slopes in plan.

The indicated physiognomic features are used as diagnostic symptoms of the origin of
slopes, the qualitative uniqueness and intensity of ancient and recent slope processes. As is
well known, Penck (1961) assigned great value to slope profiles as indicators of the relations

between intensity of tectonic uplift and denudation. He stated that convex slopes indicate ascending development of relief, but that concave and straight slopes indicate descending and steady development. Although this view of Penck's was based on incomplete data, from our point of view we may still state that a convex profile with transition from a gentle segment of slope downward to a considerably steeper and higher normally indicates recent tectonic uplift of the earth's crust and profound accelerated stream erosion caused by this uplift.

The shape of a slope also serves as an indicator of the general geographic conditions for development of the landscape. Convex and convex-concave slopes form chiefly in warm humid climates that favor chemical weathering and slope-wash and solifluctional slope processes, with which the lowering of divides and leveling of slopes are generally associated. Convex slopes are developed chiefly in semiarid and arid climates because of intense washing of detritus down steep slopes and the formation of gently inclined bahadas at the base of the slopes.

By slope shapes we may judge the geologic structure of a locality. Thus, convex and stepped slopes are observed where beds of strong rocks alternate with rocks more susceptible to weathering.

The character of ancient and recent slope processes determining the genetic types of slopes (Scheidegger, 1961) appears more directly in the physiognomic features of slopes. The natural slope angle (angle of repose) of sandy masses permits us to separate slopes subjected to rapid gravitational processes, such as collapse slopes and talus, from slopes on which slow processes of soil flow and creep have operated. On collapse-talus slopes, steep walls where material has been plucked or torn away may be observed, with the apron of collapse-talus material below. Landslide slopes are very clearly distinguished physiognomically by their characteristic cirquelike walls above and hummocky-stepped slide material below. Slopes of sheet erosion and detritus are recognized by their smooth convex-concave profile, the upper convex part corresponding to removal of material, the lower concave part to accumulation. Solifluctional slopes, especially cryosolifluctional, are recognized by specific peculiarities of the micromorphology.

The physiognomic features of slopes permit us to determine the intensity of present-day slope processes. For example, the intensity of gullying may be judged by the steepness and length of slopes and by the slope shape in profile and plan. Long steep slopes, convex in longitudinal profile, and convex in plan (i.e., with converging drainage lines) are especially subject to erosion.

The study of slopes as a statistical population with the plotting of distribution curves for slope angles or trigonometric functions permits us to shed light on such landscape features, facilitating genetic interpretation, evaluation of endogenetic and exogenetic aspects of the landscape, and geomorphological regionalization of the territory (Grachev, 1963; Molchanov, 1967).

On composite histograms of slopes plotted by Molchanov (1967) for a region in Buryat ASSR, a series of peaks are found on the distribution curve (Fig. 4). These histograms may be separated into several particular unimodal distributions characterizing homogeneous slope populations, differing from the general community not only in steepness but also origin. One may readily distinguish flood plains and stream terraces (slope angles within limits of 0-8°, characteristic angle of 3-4°), snowfield surfaces (2-13°, characteristically 7°), fan and slope-wash surfaces (8-18°, characteristically 13°), slope-wash slopes of present-day origin (18-42°, characteristically 30°), and gravity slopes (angle of repose, greater than 42°).

Relief forms are three-dimensional geometric bodies, limited by surface and framework elements. Simple and complex relief forms may be distinguished.

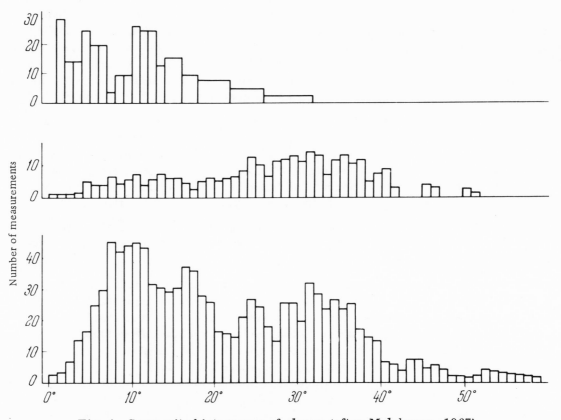

Fig. 4. Composite histograms of slopes (after Molchanov, 1967).

Simple relief forms are bounded by slopes and summit or bottom surfaces with one point or line framework element. They are divided into positive and negative, which, in turn, may be subdivided into equant and elongate.

Equant positive forms are mounds, hillocks, hills, peaks, isolated mountains. Elongated positive forms are bars, banks, ledges, benches, ridges, crests, ranges. The above forms are listed approximately in order of increasing size. (Translator's note: the English terms are not as systematic in this respect as the Russian terms.) Large forms are usually complicated by smaller forms and may be represented by simpler forms if we mentally eliminate all the surfaces complicating them. Equant and elongate negative forms are further subdivided into closed (basins, depressions) and open. Open elongate forms include valleys and canyons.

Common simple forms are characterized by approximations of various geometric bodies or by comparing them with actual objects. Thus, equant positive forms may be domal, conical, pyramidal (three-, four-sided), and, with these, complete or truncated. Equant negative forms may be saucer-shaped, bowl-like, funnel-shaped, cirque-like, and cylindrical. Elongate negative forms are trapezoidal, V-shaped or wedge-shaped, U-shaped, box-like, and so forth. All forms, in addition, are symmetrical or asymmetrical (Fig. 5).

In plan, forms are equant, slightly elongate, and markedly elongate, and the latter are subdivided into closed or connected and open or disconnected, the latter exhibiting various patterns. Actually, relief forms usually deviate from the shapes of simple bodies, and their outlines in plan prove to be more complicated. The degree and character of this divergence, defined by coefficients of approximation, are important physiognomic features of simple forms.

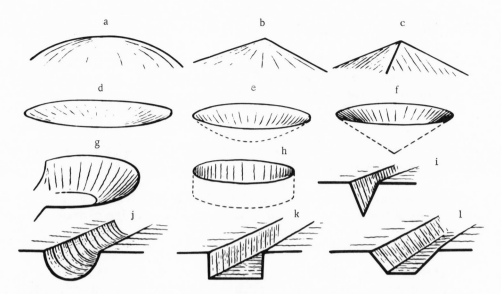

Fig. 5. Physiognomic varieties of simple relief forms. a) Domal; b) conical; c) pyramidal; d) saucer-shaped; e) bowl-shaped; f) funnel-shaped; g) cirque-like; h) cylindrical; i) V-shaped, j) U-shaped; k) box-like; l) trapezoidal.

Each relief form is defined by the following morphometric indices: absolute elevation of the crest or floor, of the crest line or thalweg (greatest, least, mean), relative height of positive and depth of negative forms, transverse dimensions, length and width, of elongate forms, trend of long axis (general, and of individual elements), directions of slope (exposure) of steep and gentle asymmetrical forms, and radius of curvature of curvilinear forms. Along with absolute values, some relative indices are also of importance: coefficient of elongation of forms, i.e., the ratio of length to width, the relief coefficient, i.e., the ratio of height (depth) to width of a form, and the coefficient of tortuosity, i.e., the ratio of true length of a form to the length measured in a straight line.

The origin of individual relief forms is clearly manifested in the physiognomic features. When observing nature or studying relief by means of detailed maps or air photos, the various erosional forms due to the activity of water are readily distinguished by specific characteristics of purely external features: gullies, canyons, valleys; forms of fluviatile origin proper are flood plains, terraces, channel bars, levees, meander scars, fans; features of groundwater are dolines, collapse sinks, poljes; forms of alpine and continental glaciation include cirques, horns, U-shaped valleys, roches moutonnées, terminal moraines, ground moraine, kames, eskers, drumlins; eolian forms include barchans, longitudinal and transverse dunes, seifs, parabolic dunes, deflation basins, and yardangs; forms of marine origin are terraces, spits, bars, tombolos, and volcanic features such as cones, craters, calderas, and so forth.

The external aspects of forms permit us not only to determine the basic factors producing them, but also to evaluate the dynamic conditions and stages of their development. Thus, the destructional, destructional—constructional, or constructional phase of development of fluvial forms is revealed in their longitudinal and transverse profiles, the direction and strength of prevailing winds in aspects of eolian forms, the direction of detrital transport in features of marine depositional forms. Structural-denudational forms (mesas, cuestas, dikes, necks, etc.) are readily recognized, the physiognomic features of which point to genetic and structural-morphological peculiarities of protective and resistant geologic bodies.

For genetic interpretation of forms we call attention to the above morphometric indicators, particularly the coefficients of elongation, relief, and tortuosity. Thus, a morphogenetic series of erosional forms from the youngest types of gullies and canyons to the most advanced valleys of plains regions may be based on the coefficient of relief. This index defines the viscosity of lava and the origin of volcanic accumulations. For example, the extrusive domes of Shiveluch, made up of viscous lavas, have a coefficient of relief of $\frac{1}{2}$ to $\frac{1}{3}$; whereas effusive basaltic cones and flows of the Klyuchi cone have a coefficient of $\frac{1}{80}$. For volcanoes of mixed origin, the coefficient of relief has an intermediate value (Menyailov, 1954).

Composite relief forms, in contrast to simple forms, are combinations of series of irregularities, which, however, do not destroy the significance of these forms as complete morphological units. They also may be positive or negative, equant or elongate, open or closed. These forms are characterized as a whole and by elements by the same physiognomic indicators discussed above. An important supplementary feature is found in the morphographic and morphometric relations of the combined forms.

Complicating details relative to form as a whole and in relation to each other may be inset or superimposed aspects. These relations may be established accurately not only by studying the geologic structure of the form, but commonly by purely physiognomic indicators as well. For this it is sometimes sufficient to plot (restore) missing surface elements, being guided in this by the principal (dominant) visible faces of relief. We may thus readily establish the existence of inset or incised gullies within larger canyons, inset or incised valleys with series of stream terraces, sinks inset in larger basins such as karst valleys, small volcanic cones within larger ones, eskers or kames set on morainic ridges or upland, parasitic cones on the slopes of basic volcanoes (Fig. 6). It is especially important to differentiate one-sided stepped or terraced forms, which may be considered physiognomically abutting features.

Study of the relations of elements in complex rugged relief reveals a sequence in their development, since superimposed, inset, or abutting forms are younger than the basic forms. Even where a composite form develops as a result of a single landscape-forming process (such as a morainic ridge, complex hills, composite barchans), its details arise during the latest stage of this process and are usually the most dynamic.

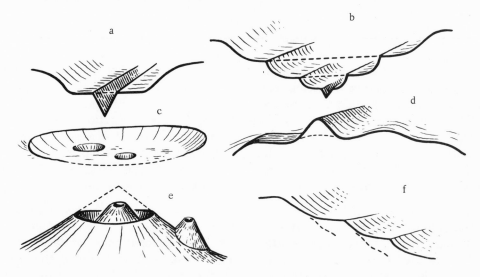

Fig. 6. Some varieties of morphological relations of forms. a) Incised canyon; b) incised (terraced) valley; c) sinks on floor of depression; d) esker, superimposed on morainic hills; e) inset and superposed small volcanic cones (central and adventive); f) one-sided (abutting) terraces.

The structure of composite forms is characterized also by relations of the morphometric indices of its elements: length, width, height (depth), slope angle, longitudinal inclination of framework lines (especially thalwegs). These relations shed light on the dependence of small forms on larger or on the structure of the relief-forming process. Thus, outward aspect, depth, and length of a gulley on a valley floor depends on the form, size, and longitudinal inclination of the valley; the relation of size of elemental and composite barchans with other small and large sand forms reveals the structure of air currents leading to the formation of the composite eolian landforms in deserts.

Groups of relief forms may be considered in two plans: as groups of simple one-dimensional forms (small, medium, and large) and as groups or combinations of all forms (simple and composite, small and large) found in a given region and forming the landscape as a whole. Here we are speaking of such landscape forms as are found adjacent to each other and form systematic spatial combinations (associations). The name of the group comes from the principal positive and negative forms in the composition of the group.

The physiognomic characteristics of groups are summed up from the above generalized morphographic and morphometric characteristics of the form components. Important supplementary characteristics are the aspects of spatial distribution of forms within a given group.

The disposition of individual equant forms may be individual or nested, disordered or ordered, ordered in the form of chains, elongated in rectilinear or curvilinear fashion, in one or two or more mutually intersecting directions. The design of elongate forms is well represented by crestal and thalweg framework lines.

The distribution of forms in plan is also characterized by quantitative indices. Primarily this means density of forms, defined by the number per unit area or by the distance between them. In particular, for the density index of the drainage network, the length of thalwegs per square kilometer or the average distance between adjacent thalwegs is adopted. In addition, it is recommended that the dimensions and relations of areas occupied by the different forms within a given group be determined, such as hills and depressions, ridges and valleys. The trend of elongated chains of forms, valleys and divides, exposures of steep and gentle slopes of asymmetrical forms are marked with appropriate numerical indices.

Associations of forms are distinguished by special patterns of their faces, a description of which may be obtained by the methods indicated above (approximation of geometric figures, determination of tortuosity, linear dimensions, and areas of mapped feature).

On going from a study merely of individual relief forms to a study of groups, broad possibilities are opened up for using the comparative method of investigation, utilizing cartographic and variation-statistical methods, which permit us to attain a deeper perception of landscape, even when working only with some single physiognomic features. The isolation of just a single typically recurring relief form, possessing definite outward aspect, permits us to draw important genetic conclusions (Shchukin, 1964). Thus, ridge—valley relief very definitely indicates a dominant role of erosion by water and by slope processes. Landscape with ridges, ridges and closed depressions, or just closed depressions characterize eolian reworking of sands in the desert. Irregular glacial deposition leads to hummocks and basins or hummocks and ridges.

It is possible to establish paragenetic complexes of forms possessing strictly individual physiognomic features, facilitating genetic interpretation of the landscape. A classic example is the complex that forms at the edge of a glacier, consisting of terminal moraines, adjoining outwash plain on the front and hilly ground moraine behind with flat floors of lakes now drained by breaching of their dams.

The outward aspect of relief is characterized also by the distribution of its morphographic and morphometric features along the vertical. As we know, on plains and, especially, in mountains, the shapes of the surface change smoothly or very sharply with height. This is manifested in morphological stages and belts.

Stages of relief are expressed in sudden increments of absolute and relative height, which may be found at several hypsometric levels (stages), forming a systematic height-group of summits, remnant surfaces, terraces, benches, slope inflections. The required study of physiognomic features of landscape stages supplies abundant, although by no means exhaustive, material for genetic construction. By the outward aspect of smooth and steeply sloped forms, by the distinctive features of their groups and distributions, it is possible to draw preliminary conclusions concerning the passive role of complicating structural–lithic conditions in landscape development, concerning the decisive effect of recent tectonic movements causing alternating rising and descending modifications of the landscape and leading to the formation of several levels of denudation surfaces, differing in age, and concerning the role of faulting.

Morphologic zones appear in mountains, where they reflect vertical climatic belts (Shchukin, 1964). In alpine mountains the uppermost zone, with serrated crests and sharp pyramidal peaks, is sharply isolated, bound on all sides by cirques. Below this we find a belt with comparatively gentle slopes and subdued relief features, created in the past by the action of firn and ice, and now occupied chiefly by alpine and subalpine meadows. Still farther downslope appear steep-sloped valleys, formed chiefly by profound stream erosion and usually covered by mountainous forests in the humid climate of middle latitudes.

For more distinct manifestation of some physiognomic features of the earth's surface, it is expedient to prepare morphographic and morphometric maps, cartograms, and diagrams (Spiridonov, 1952). Genetic interpretation of the landscape (explanation of the role of ancient structures and recent tectonics in development of the landscape, its exogenetic relations, and stage of development) is greatly facilitated by cartograms of relief energy, maps of intensity of horizontal and vertical dissection, slope angles, structural contours on the erosion surface, structural contours on higher surfaces (generalized), and other aids. Soviet investigators have accumulated a great amount of experience in interpreting all these maps for structural-geomorphological investigations and in prospecting for oil and gas structures.

For a generalized morphometric description of groups of forms we have recourse to the plotting of distribution curves of absolute and relative height, slopes of surfaces, dimensions of forms in plan (length and width) with an expression of median and modal values, asymmetry, and other indices of distribution. As Soviet and foreign investigators have shown (Chichagov and Devdariani, 1963; and others), such methods permit one to establish statistical patterns by which one morphological complex may be distinguished from another, and consequently, may be used for diagnostic purposes.

On histograms of height distribution, plotted by Sharapov (1967), deeply dissected alpine relief is characterized by uniform (amodal) distribution (Fig. 7a); mesas separated by broad valleys exhibit antimodal (U-shaped) distribution (Fig. 7b); lowlands with few hills, right skewed amodal distribution (Fig. 7c); plateaus cut by canyons, left skewed amodal distribution (Fig. 7d); mature mountainous terrain of intermediate height, symmetrical unimodal distribution (Fig. 7e); low-mountain remnants and small peaks, right-skewed unimodal distribution (Fig. 7f); strongly dissected mountains or plains, left-skewed unimodal distribution (Fig. 7g); stagewise relief with a series of planation surfaces, polymodal distribution (Fig. 7h); and dissected mesas and plateaus, amodal distribution with a secondary peak (Fig. 7i).

On the basis of statistical treatment of the trends of framework lines in the landscape (thalwegs, ridge crests) or chains of individual equant forms (such as sinks), diagrams and

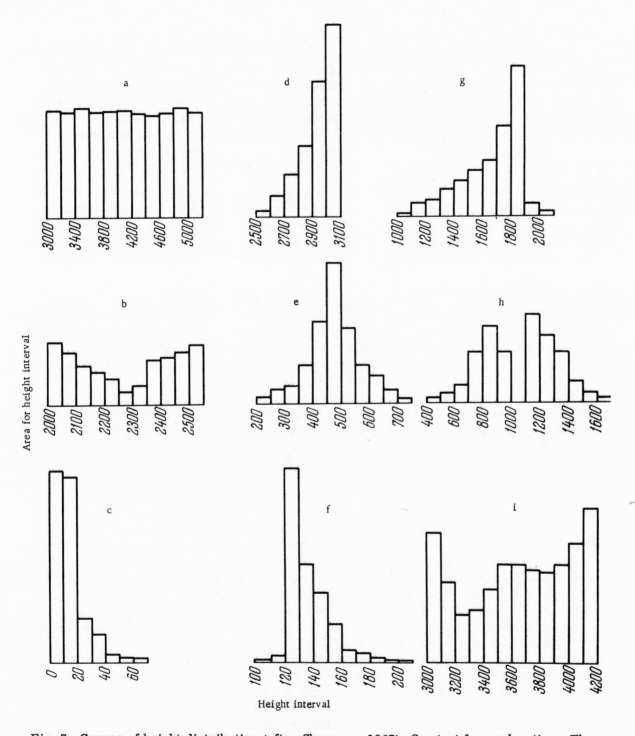

Fig. 7. Curves of height distribution (after Sharapov, 1967). See text for explanation. The vertical coordinate on the histograms represents area occupied by surfaces at the indicated interval.

map patterns are constructed: rose diagrams. Such diagrams graphically represent systems of joints, faults, and other structural elements having an important effect on the relief pattern in plan view.

For broad genetic generalizations, considerable value is found in harmonic analysis of the relief, with clarification of the forms of different orders (dimensions). As shown by Piotrovskii (1964), relief forms are systematically grouped into 18 orders from smallest to largest (global scale) with constant relations of length, width, and height (depth), and to this morphometric series there corresponds a series of tectonic structures. With extensive systematic collection of morphometric data, investigations may be organized for clarifying the correlation between the various quantitative indices, such as between height and area of catchment areas, between exposure and steepness of slopes, and so forth (Horton, 1945; Filosofov, 1960; Borsuk et al., 1966).

In conclusion we must note that the indicated component parts of the landscape are necessarily considered completely, as a whole, and not in isolation from each other. Their mutual interrelations and combinations must be taken into account. Only in this way will the outward aspect of the landscape be perceived not as some random chaotic accumulation of forms but as a systematic morphological complex, having a definite origin and development.

LITERATURE CITED

Borsuk, O. A., Gorunova, M. A., Likhacheva, É. A., and Simonov, Yu. G., Morphometric Characteristics of the Landscape and Their Analysis in Geomorphological Investigations, Questions on the Geology of the Baikal and Trans-Baikal Regions (Vopr. Geologii Pribaikal'ya i Zabaikal'ya), No. 1 (3), Chita (1966).

Chichagov, V. P., and Devdariani, A. S., "Morphometry in the works of H. Baulig (France) and A. N. Strahler (U.S.A.)," in: Questions of Geography, Coll. 63, Moscow (1963).

Devdariani, A. S., "Quantitative methods in studying relief," in Questions of Geography, Coll. 63, Moscow (1963).

Devdariani, A. S., "Mathematical methods," Itogi Nauki, Seriya Geogr., No. 1, VINITI, Moscow (1966).

Devdariani, A. S., Mathematical Analysis in Geomorphology, Nedra, Moscow (1967).

Efremov, Yu. K., "Experiment in morphographic classification of elements and simple forms of the landscape," in: Questions of Geography, Coll. 11, Moscow (1949).

Filosofov, V. P., A Brief Guide to the Morphometric Method of Searching for Tectonic Structures, Izd. Saratovsk. Univ. (1960).

Gerenchuk, K. I., Theoretical Patterns in Orographic Features and Drainage Networks of the Russian Plain, Izd. L'vovsk. Univ. (1960).

Grachev, A. F., "Use of the quantitative method in morphometric analysis of slopes (in connection with the search for tectonic structures)," in: The Morphometric Method in Geological Investigations, Izd. Saratovsk. Univ. (1963).

Gvin, V. Ya., "Use of morphometry in structural investigations of the upper and middle Volga region and the Kama region," in: Questions of Geography, Coll. 63, Moscow (1963).

Horton, R. E., "Erosional development of streams and their drainage basins; hydrophysical approach to quantitative morphology," Geol. Soc. Am., Bull. 56 (1945).

Khain, V. E., General Geotectonics, Nedra, Moscow (1964).

Makkaveev, N. I., The Stream Bed and Erosion in Its Basin, Izd. AN SSSR, Moscow (1955).

Menyailov, A. A., "Basic stages in development of the Shiveluch Volcano," Trudy Labor. Vulkanol., AN SSSR, No. 8 (1954).

Miller, V. C., and Miller, C. F., Photogeology, McGraw-Hill, New York (1961).

Molchanov, A. K., "Characteristic and limiting angles of slope in the southern part of the Buryat ASSR," in: Methods of Geomorphological Investigation, Vol. 1, Nauka (Siberian Branch), Novosibirsk (1967).

Nikolaev, N. I., Quaternary Tectonics and Its Expression in the Structure and Relief of the SSSR, Gosgeoltekhizdat, Moscow (1962).

Penck, W., Morphological Analysis [Russian translation from German of the 1924 edition], Geografgiz (1961).

Petrusevich, M. N., Aerial Methods in Geological Investigations, Gosgeoltekhizdat, Moscow (1962).

Piotrovskii, V. V., "Use of morphometry for studying relief and structure of the earth," in: The Earth in the Universe, Mysl', Moscow (1964).

Piotrovskii, V. V., "The third conference on mathematical methods in geomorphology," Izv. Akad. Nauk SSSR, Seriya Geogr., No. 1 (1965).

Scheidegger, A. E., Theoretical Geomorphology, Prentice-Hall, Englewood Cliffs, New Jersey (1961).

Sharapov, I. P., Functions of the Height Distribution of Relief. Relief and Mathematics, Mysl', Moscow (1967).

Shchukin, I. S., General Geomorphology, Vols. 1 and 2, Izd. MGU (1960) (1964).

Spiridonov, A. I., Geomorphological Mapping, Geografgiz, Moscow (1952).

Spiridonov, A. I., "The concept of 'relief types,'" in: Questions of Geography, Coll. 46, Moscow (1959).

Tskhovrebashvili, Sh. A., "Determination of the number of basic stages of denudational surfaces in mountainous terrain," in: Problems of Denudational Surfaces, Nauka, Moscow (1964).

Volkov, N. G., "The method of tectonic analysis of longitudinal stream profiles," Izv. Akad. SSSR, Seriya Geogr., No. 2 (1964).

INTERPRETATION OF LANDSCAPE AS AN INDICATOR OF GEOLOGIC STRUCTURE

I. S. Gudilin

The study of landscape is involved in the complex of geologic exploration and surveying. One of the basic objectives in this is clarification of the interrelations between elements of relief and geologic structure. If we consider the elements of geologic structure to be possible indicated objects and the landscape elements to be the indicators, then we may readily conclude that in geomorphological and geologic investigations, the methods of indicator geomorphology should be widely applied.

Indicator geomorphology is widely used in various fields of the geologic and geographic sciences (soil science, plant geography, geology, engineering geology, hydrogeology, and still others). Whereas problems of indicator plant geography and indicator landscape science have been the subjects of a number of special investigations (Viktorov and Vostokova, 1963; Viktorov, 1967; Vinogradov, 1957; Chikishev, 1960), problems of indicator geomorphology have been considered only incidentally in various works (Miroshnichenko, 1966; Lungersgauzen, 1964; Kobets, 1964; Gudilin, 1967).

Geomorphic indicators are widely used in geological surveys, especially in shedding light on elements of tectonics, mapping unconsolidated Quaternary deposits, studying present-day relief-forming processes, determining composition of rocks, and investigating many mineral deposits (Lungersgauzen, 1964). The use of geomorphic indicators furnishes especially good results in geomorphic interpretation of air photos. Such interpretation is used both in complex geologic exploration as well as in independent geomorphological surveys, appreciably increasing the quality and reducing the time of field work in both geology and geomorphology.

Even where the geologic structure of a region is complex, geomorphic interpretation permits us to find geomorphic indicators of structural elements. Geologic interpretation therefore reduces in part to geologic interpretation of the results of geomorphic interpretation. It is difficult to draw a boundary between geologic and geomorphic interpretation, especially in geologic exploration of regions with depositional and structural relief forms. Geomorphic indicators in geomorphic interpretation permit us to establish the origin of the landscape, solve a number of problems relative to endogenetic and exogenetic relief-forming processes, their trends and intensities, determine the relative age of the landscape, and more.

Geomorphic interpretation, made in the complex of geomorphic and geologic field investigations, should be considered one of the most productive methods of indicator geomorphology. In this sense geomorphic interpretation is "indicator" interpretation. It is from this point of view of indicator properties of the landscape that we consider geomorphic interpretation in the present article.

The possible use of air photos for establishing the indicator role of landscape elements depends primarily on the dimensions of these elements and the scale of the air photos. Indicator geomorphology is successfully used for geologic indication in microrelief forms that appear on large-scale air photos.

In geomorphic interpretation of medium- and large-scale air photos, important details of relief features may be obtained, having indicator significance, that may find no reflection even on large-scale topographic maps (terrace edges, karst and thermo-karst forms, collapse features, and various gravitational, eolian, and other erosional forms). The possibility is especially good for finding such features on small-scale photos, perspective air photos, high-altitude (satellite) photos.

The indicator role of landscape in geomorphic interpretation for geologic purposes is very great. Geomorphic interpretation permits us to establish the relationship between landscape and rocks, and fold structures. The types and forms of landscape are important indicators of recent tectonic structures, such as large anticlinal uplifts, horsts and grabens, faults, and other features.

In geomorphic interpretation it is possible to establish an indicator role of present landscape elements for relict and buried landscape elements. Good results may always be obtained in the geomorphic interpretation of ancient erosion surfaces. Relics of these surfaces are very difficult to study by ordinary methods of investigation, whereas the use of small-scale air photos (and photo composites) in combination with large-scale and perspective photos permit one to establish (with ground checking, of course) the role of endogenetic processes in the dissection and deformation of these surfaces.

Especially valuable conclusions may be obtained in geomorphic indication interpretation during prospecting for local oil and gas structures (Western Siberia, Taimyr depression, Belorussia, Sakhalin, and elsewhere), prospecting for gold placers (Far East, Yakutia, Tuva, and elsewhere), and other exploration.

Indicator study of landscape from air photos is possible only on the basis of a wide use of interrelations between all natural components. This approach to geomorphological (and not only geomorphological) interpretation is called the landscape method of interpretation. This method is based on a study of the interrelations between direct interpretation of natural components (forms and types of landscape, characteristic hydrographic networks, vegetation, elements of man's activity) and geologic structure, groundwater, and other "internal" components. Study of the interrelations is made within a landscape and its morphological parts.

For complete use of the landscape method of interpretation and clarification of the indicator role of the landscape for elements of geologic structure, it is necessary to consider the possibilities of this method under different natural conditions: bioclimatic and structural-tectonic. This follows from the distribution pattern of bioclimatic and geologic–geomorphic groups of landscape components. Interpretation always reduces to recognition, within a landscape or its morphological parts, directly interpretable natural components: relief, drainage patterns, vegetation, activity of man. These directly interpretable components create definite designs in the images on air photos (structure and texture), the study of which plays an important role in interpreting geomorphic and geologic features of the region.

The essence of the interpretation of landscape as an indicator of elements of geologic structure follows from the very content of an air photo, on which the outer components are depicted, forming "physiographic complexes": regional combinations of relief, drainage patterns, vegetation, traces of man's activity. Physiographic complexes may contain any combinations of the indicated components, the most important of which (and the most physiognomic) is relief. The combinations (complexes) of relief and drainage patterns, and of relief and vegetation, are the most reliable indicators of geologic objects.

In interpreting the landscape for geologic purposes, the same physiographic complexes appear that are interrelated with elements of geologic mapping and are distributed only within those elements. In turn, for landscape elements, indicator complexes may be combinations of plant groups with hydrographic elements, traces of man's activity, and groups of physiognomic components.

Indicator complexes, the most important component in each of which is relief, form a characteristic structure of the pattern in areas of the geologic object. According to the scale of the geologic investigations and the scale of satellite or rocket photos, photo composites, photomaps, or air photos, we may distinguish indicator complexes corresponding to landscape boundaries, localities, and districts, within which there also exists an interrelation between the physiographic complexes and elements of the geologic structure.

Complexes indicating geologic objects are indicator complexes (at a locality) or complex interpretative features. Natural components – relief, water surfaces, plant cover, results of man's activity – are particular indicators (at a locality) or interpretive features of objects. Particular and complex interpretive features may refer to so-called indirect interpretive features. Complex indicators may be used as interpretive features only when the scale of the air photos permits one to recognize them by so-called direct interpretive features.

Direct interpretive features include the size and shape of the object, its position in the locality relative to other objects, the intrinsic and incident shadows, tones, and colors of the image, the image design (structure and texture). Complex and particular indicators conform in their distribution to the structural-tectonic and bioclimatic patterns. In this connection, when interpreting types and forms of relief, unconsolidated Quaternary deposits, or bed rock, we should use only those indirect (particular and complex) interpretive features that are characteristic of actual natural conditions of the investigated region. In this sense we may speak of intrinsic interpretive features of geologic and geomorphic objects inherent in the objects by virtue of their characteristic internal qualities (nature of jointing in granites, form of flood plain of a river, alluvial fan, etc.) and we may speak of zonal interpretive features. The latter features of geologic and geomorphic objects are associated with bioclimatic peculiarities of zones (subzones), vertical vegetation zones, within which specific communities are developed, or recent exogenetic processes, man-made or animal-caused complexes.

The possibilities of geomorphic interpretation made during exploratory geologic work are determined to a considerable extent by peculiarities of the plant cover, traces of man's activities, characteristic drainage pattern, morphostructural and morphosculptural features.

Vegetation is an important factor, substantially affecting the possible use of landscape as an indicator of geologic structure. In some cases the plant cover accentuates the role of relief, as it were, as an indicator of the elements of geologic structure. In other cases, on the contrary, it worsens the possibility of interpretation. The role of vegetation as an indicator of geologic structure has been greatly clarified by geobotanical and geological investigations (Viktorov and Vostokova, 1963; Vinogradov, 1957; Kobets, 1964; Shvyryaeva, 1964).

Geobotanical features due to differences in the plant cover and to the image of vegetation on air photos are used as indicators of geologic structure. Vegetation either stands out independently or, more frequently, in a complex with landscape elements (microrelief and meso-microrelief), forming complex indicators, creating characteristic image patterns on the air photos.

In the tundra zone, types of tundra are combinations of microrelief forms with vegetation and soil (banded, tussocky marsh, polygonal tundra, and others) and as complex indicators are commonly indicators of elements of geologic structure (Gudilin and Konstantinova, 1965). In the taiga and forest zones, the plant cover most commonly lessens the possibility of using landscape as an indicator for geomorphic or geologic interpretation, especially in regions of bedrock exposures and complex tectonic structure. Much more rarely, combinations of micro- and mesorelief forms with different types of wooded and shrubby vegetation permit us to identify relief elements, on the one hand, and to interpret related elements of geologic structure. Different types of swamps, closely associated in these zones with mesorelief forms and composition of the underlying unconsolidated rock, are also important interpretive features of geologic structure. In the zone of wooded steppes, shallow depressions undiscernible at the site are clearly distinguished on air photos because of high moisture content in the soil and the presence of variherbaceous—meadow, shrubby, and other vegetation. In deserts and semideserts, low areas may be readily identified by light patches of takyrs (clay desert zones) and solonchaks (dry salt lakes), readily interpretable on air photos.

Thus, vegetation may be employed more reliably in geomorphic and geologic interpretations when it forms complex indicators with landscape elements.

Traces of man's activity — "cultural landscape" — the character of inhabited localities, agricultural areas, traces of reclamation efforts, paths and roads, working of structural materials, hydroelectric installations, and so forth — are usually related closely to landscape elements and to peculiarities of geologic structure, and this permits us to use these features as indicators. Inhabited localities and cities are generally found on certain elements of mesorelief most favorable for these purposes (high terraces at the mouths of rivers, steep banks, alluvial fans in arid regions, drained divide areas, upper parts of morainic hills, etc.). Paths and roads permit us to judge of the slope of the surface and the nature of the underlying rock. The irregular width and indistinctness of railroad beds are characteristic of sand on plains. In such uneven districts, where moisture is in excess, loams and clays are responsible for railroad tracks going around. On divides, such deposits are characterized by clear boundaries and constant width of the railroad beds (Kobets, 1964). Swampy tracts in zones of low relief are characterized by dark bands of the road network with many detours. In mountainous regions paths usually go along well-drained and warmed elements of the landscape. Changes in direction and beds usually indicate steepness of slopes. Thus, traces of man's activity emerge as indicators of landscape and elements of geologic structure. Mostly they should be considered with landscape elements as complex indicators.

The character of the drainage pattern is also usually closely related to relief and geologic structure and is used as a particular indicator for these components. The hydrographic elements form with relief complex indicators of units for geologic mapping.

The hydrographic elements in great measure determine the design of the photo image of morphogenetic types and forms of landscape. Important indicators of relief and structure, including rock properties, Quaternary structures (trend, intensity, and amount of movement), and thickness of unconsolidated deposits are drainage patterns, density of drainage channels, orientation, depth and intensity of incision, character of transverse and longitudinal profiles of the valley, configurations of lakes, types and developmental trend of swamps, and still others. For geomorphic and geologic interpretation, classifications of drainage patterns are used (Petrusevich, 1961).

Drainage patterns are commonly classified by the character of their forms and disposition of tributaries in the overall system of a given part of the basin. The form of the pattern is compared with forms of well-known objects. The classification of drainage patterns proposed by Parvis (1952) contains six basic patterns and numerous modifications. Despite the great variety of proposed patterns, the classification includes types having no generally direct relation to indication of geologic structure (such as the pattern of drainage tubes). Below we have presented characteristic patterns, using the above-mentioned classification, with modifications. The following types of patterns are differentiated: dendritic, parallel, radial, structural, unsystematic, and subtypes of these (Fig. 1).

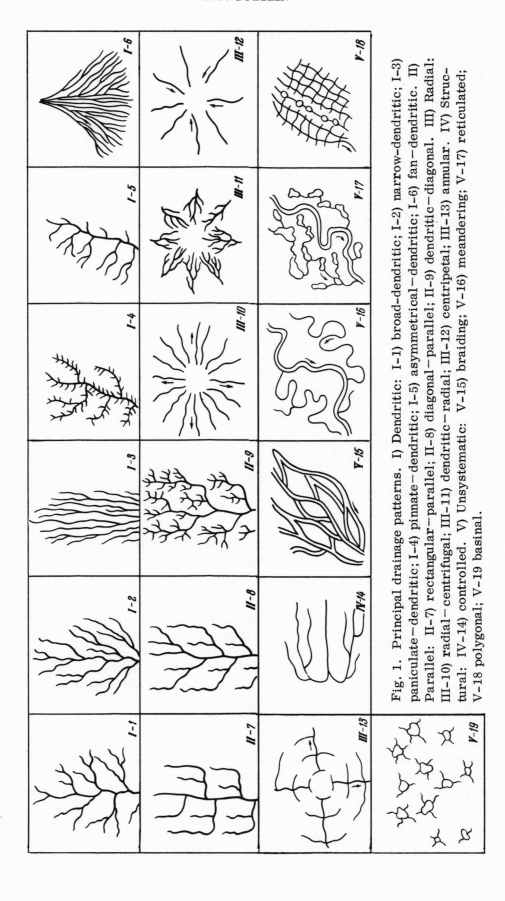

Fig. 1. Principal drainage patterns. I) Dendritic: I-1) broad-dendritic; I-2) narrow-dendritic; I-3) paniculate-dendritic; I-4) pinnate-dendritic; I-5) asymmetrical-dendritic; I-6) fan-dendritic. II) Parallel: II-7) rectangular-parallel; II-8) diagonal-parallel; II-9) dendritic-diagonal. III) Radial: III-10) radial-centrifugal; III-11) dendritic-radial; III-12) centripetal; III-13) annular. IV) Structural: IV-14) controlled. V) Unsystematic: V-15) braiding; V-16) meandering; V-17) reticulated; V-18 polygonal; V-19 basinal.

Dendritic drainage patterns resemble the branches of a tree. In the nature of the branching, the width of the system, the angle at which lateral valleys join the main valley, and orientation, six subtypes are distinguished. The broad—dendritic pattern (Fig. 1, I-1) is an indicator of lithically and structurally homogeneous rocks in very low relief. It is characteristic of old erosional surfaces of plains and plateaus on metamorphic and magmatic rocks (outside zones of tectonic crushing), structural plains and plateaus composed of horizontal homogeneous rocks of considerable thickness, and depositional plains composed of homogeneous sequences of unconsolidated sediments. The narrow—dendritic (subparallel) patterns (I-2) is characteristic of originally inclined plains (such as marine or lacustrine plains), belts of reclaimed marine plains. The paniculate—dendritic pattern (I-3) indicates homogeneous, rather dense rocks, uniformly dipping in one direction at a low angle. The pinnate—dendritic pattern (I-4) is characteristic of such homogeneous rocks as loess, loessial loams, and clay. The asymmetric—dendritic pattern (I-5) indicates dense, relatively homogeneous rocks with homoclinal structure (cuestal ridges). The fan—dendritic pattern (I-6) points to alluvial fans and indicates composition of rock making up the fan, climatic conditions, and recent tectonic activity in the locality.

Parallel drainage patterns are characteristic of a rectilinear main valley and rectilinear lateral tributaries, forming a network of parallel segments that join at various angles. This pattern underscores the structural peculiarities of the region. Three subtypes are recognized. The rectangular—parallel pattern (II-7) is characteristic of linear structures with homoclinal or steeply inclined beds. It is also characteristic of intersecting systems of joints or faults. The diagonal—parallel pattern (II-8) is characteristic of homoclinal or nearly horizontal beds and jointed volcanic rocks (small lateral tributaries are controlled by the strike of the structure, and the angle at which they join the main stream is controlled by inclination of the fold axis). The dendritic—diagonal pattern (II-9) is a variant of the rectilinear pattern, but differs in the branching of secondary tributaries. It is characteristic of homoclinal inhomogeneous rocks, and also of homogeneous rocks cut by large faults, where the main rectilinear net is controlled by structure, but the secondary dendritic pattern commonly develops on a mantle of unconsolidated sediments.

Radial drainage patterns characterize the radial arrangement of valleys, and are typical of domal uplifts, slopes of volcanoes, large basins, local structures, and closed depressions. Three subtypes are distinguished. The radial—centrifugal pattern (III-10) is characteristic of domal uplifts, local structures, morainic hills, slopes of volcanoes. The dendritic—radial pattern (III-11) is characteristic of the previously mentioned elements, but with advanced stages of development of the drainage network. The centripetal pattern (III-12) indicates closed basins; with more advanced stages of development, a variety of patterns may appear. The annular pattern (III-13) forms when annular tributaries develop and drain into radial main valleys. It is characteristic of eroded domes or basins in which concentric or arcuate ridges are sharply expressed.

Structurally controlled drainage patterns (IV-14) are found where valleys follow the trend of folds (being controlled by them). They are hence characteristic of folded regions with well-defined expression of the folds in the landscape. The pattern may approximate the parallel pattern in appearance.

Unsystematic drainage patterns represent an absence of any well-defined valley system. It is most typically found on depositional plains in regions of steady subsidence, but it occurs also on karst surfaces, in regions where collapse features are common, and elsewhere. Five subtypes of this pattern are recognized. The braided pattern (V-15) is characteristic of alluvial plains, flood plains, large broad valleys of streams that have reached the equilibrium profile. The meandering pattern (V-16) is characteristic of alluvial and lacustrine—alluvial plains in the stage of valley damming and swamp development. The reticulated pattern (V-17) is confined to hilly morains, wooded herbaceous swamps, and lacustrine, alluvial, and lacustrine–alluvial plains. The polygonal pattern (V-18) is characteristic of lacustrine–alluvial and alluvial–marine plains in the zone of prolonged permafrost with wide development of thermokarst. The basinal pattern (V-19) is confined to porous underlying rocks: limestones, settling sequences, and the like.

The patterns discussed above do not exhaust all the possible variants, but they give us a view of the principal types. The classification of drainage patterns is of great practical value, not only for indicator studies, but also for developing a classification of textures and structures of the photographic image and for the formalization of interpretive features. This is also the trend for possible automation of the interpretive process of air photos.

Morphostructural and Morphosculptural Features. The interpretation of external components – plant cover, traces of man's activity, hydrographic elements (as indicators of geologic structure) – may be made only in combination with interpretation of the landscape, which is a very essential component of complex indicators.

We open up the greatest possibility of using geomorphic and complex indicators in interpreting geologic structure when we consider the distributional patterns of these indicators as they depend on morphostructural and morphosculptural features of the region (Gerasimov, 1959); Meshcheryakov, 1965; Korzhuev, 1966).

The interpretation and clarification of indicator significance of landscape elements requires profound analysis of the interrelations of endogenetic and exogenetic relief-forming processes. The distributional patterns, the trends of development (dynamics), and the interpretive features of the landscape depend on endogenetic processes to a great extent.

Endogenetic processes are manifested in the external aspect of the landscape. However, in regions of intense Quaternary uplift, steeply sloped mountainous terrain having great energy of drainage and vigorous destruction and denudation of rocks, differences in resistance to denudation processes are smoothed out and indicator methods become difficult.

Indicator possibilities are also restricted in slightly uplifted or slightly dissected blocks, where denudation energy is low and bedrock is masked by a mantle of unconsolidated material. Such conditions prevail, for example, in the region of the differentially subsiding Central Amur depression. Here, in a zone of low mountains, no indicator role of landscapes is found for discriminating units in the sequence of Jurassic and Cretaceous sedimentary rocks by examination of air photos. The low mountains are covered by a mantle of unconsolidated deposits, swampy areas, and larch forests, which also make interpretation difficult.

The composition of bed rocks and the resistance of these rocks to exogenetic processes also determine in great measure the possible development of geomorphic indicators of geologic structure. Other conditions being equal, the more strongly the rocks differ in their resistance to erosion, the better landscape elements as indicators are interpreted on air photos. In regions of bedded rocks that differ in resistance to erosion, characteristic microrelief forms appear, and a characteristic pattern appears on the air photos. Most information of indicator elements in the landscape is obtained in zones of stratified sedimentary rocks, the least in beds of massive sedimentary rocks.

The amount of information obtained from photos during geomorphic interpretation for geological purposes, other conditions remaining the same, is less for metamorphic rocks than for igneous.

When rocks are horizontal or almost so, indicator possibilities of landscape elements are quite different from those for terranes of complexly folded rocks.

Fig. 2. Sculptured block mountains, in region of moderate recent uplift. Erosional pattern of morphosculpture: alternation of eroded domal divides and shallow incised valleys; absence of or weakly expressed orientation of drainage pattern and divides according to structural elements. Region is composed of Proterozoic hornblende–biotite gneisses, which have been intensely crushed and broken by a system of faults. Banding in the photo is due to alternations of patches of block placers of light tone on dense rocks and of dark bands on fractured and water-soaked rocks with moss-lichen and shrubby vegetation.

Horizontal or slightly disturbed strata are characterized by clearly defined structural elements in the landscape that give rise to relatively constant interpretive features of geologic structure over large regions.

In complexly folded regions, the conditions of geomorphic interpretation are much worse because of steeply dipping beds, the extensive occurrence of faults, and possible metamorphism of the rocks.

Each of the factors considered (status of recent tectonics in the region, composition of bedrock, expression of fold structures in the landscape), having important independent significance, nevertheless usually appears in combination with the others within the morphostructures.

Fig. 3. Structural–sculptural, fold-block mountains in a zone of moderate uplift; erosional type of morphosculpture in the arid zone. The photo clearly shows the plunging nose of an anticline and the northwestern limb of a fold composed of various Albian to Neogene rocks. The series of alternating beds of different colors and compositions (sandstone, clay, marl, siltstone) form a series of parallel ridges. The divide zone corresponds to the core of the anticline, formed on glauconitic sandstones.

Fig. 4. Structural–sculptural, fold-block mountains in a zone of moderate uplift; erosional type of morphosculpture in the taiga zone. The photo shows clearly the limb of a large anticline (dips of 15-30°). The mountains consist of Upper Cretaceous sandstones, forming massive cuestas overgrown with spruce–fir forests and dwarf pine thickets. Intercuestal lowlands are on Upper Cretaceous siltstones, and are covered by spruce–pine forests.

The morphostructures unite regions having definite trends, magnitude and rate of recent tectonic movements, and predominance of ancient structural elements for a given morphostructure, as well as groups of rock formations that are found expressed in the totality of landscape elements. This permits us to use a set of geomorphic indicators characteristic for a given type of morphostructure in making interpretations.

Let us consider some examples of morphostructures.

Block mountains (massive), sculptured, in zones of moderate recent uplift, on crystalline, metamorphic, rarely sedimentary rocks, are characterized by great complexity in the interrelations between landscape and elements of ancient fold structures. In indicator studies, the forms of relief give little information concerning elements of geologic structure. Complex interpretive features are used (Fig. 2).

Fig. 5. Volcanic plateau with a series of Holocene volcanic cones arranged along a fault zone. One may clearly see the eruptive centers and the plateau surface with characteristic forms of lava flows.

Structural–sculptural, fold-block mountains, in zones of recent uplift, composed chiefly of sedimentary rocks, are characterized by the clearest relations between landscape elements and geologic structure. In interpretations by means of geomorphic indicators, all landscape zones offer possibilities of finding a significant number of elements of geologic structure (Figs. 3 and 4).

On plateaus of chiefly volcanic rocks in the zone of recent uplift, one may readily distinguish forms associated with volcanic activity, and in eroded segments structural forms of relief may be clearly recognized (Fig. 5).

For plateaus in structural zones of moderate recent uplift, composed of horizontal or gently inclined sedimentary rocks, we characteristically find relief determined by the structure of the beds, but little changed by the disruptive effect of exogenetic processes. In eroded segments at the margins of the structural plateau, one may readily find areas and benches marking horizons with parting or cleavage along the bedding plane (as in the Ust-Urt Plateau).

On depositional plains in zones of relative subsidence or slow uplift, with a thick cover of unconsolidated deposits, the morphostructures of subsidence are accompanied by meandering streams and development of swamps (Fig. 6). Among morphostructures we may recognize manifestations of Quaternary tectonic movements of basement rocks, genetic types of unconsolidated deposits, and other features.

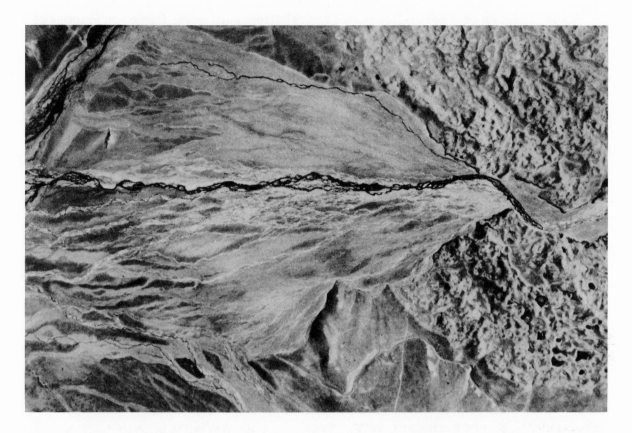

Fig. 6. Intermontane depositional plain in a zone of relative subsidence (in an alpine zone).
Two types of morphosculptural features are present: glacial accumulations (terminal-moraine
ridge) and outwash type of morphostructures.

Exogenetic processes lead to destruction of morphostructures and to accumulation of unconsolidated material, as a result of which different types of morphosculptural features are
created, in certain stages of the landscape forming morphogenetic types of relief.

Relief types under certain bioclimatic conditions determine landscape formation. For
each landscape—climatic zone and vertical belt, there is a set of typical zonal relief-forming
exogenetic processes and corresponding relief forms with specific interpretive features of
these forms.

The zonation of relief forms is subordinate to recent geographic conditions in lesser degree than the zonation of soils and plants. The relief is more "conservative" than the soil-
plant cover, since relief forms reflect not only recent climatic conditions but also ancient conditions no longer in existence. The zonation of morphosculptural features is more complex because of the presence of relict forms and because of the greater effect of morphostructural
features, which are manifested in the drainage pattern, in the depth of dissection, and so forth.

The leading exogenetic processes of destruction lead to the formation among morphostructures of the following principal types of morphosculptural features: snowbank—glacial,
exarational, erosional, denudational, arid—denudational, and more. Accumulation processes
lead to the formation of the following types of morphosculptural features: alluvial, lacustrine,
marine, slope-wash, glacial and glaciofluviatile, eolian, thermokarst, and still others.

In each morphogenetic type of relief there are specific peculiarities in indicating relief
forms for elements of geologic structure under different landscape conditions.

Low-mountain erosional relief in zones of relatively slow uplift, formed on granites of approximately a single composition and age in the various segments, will have different interpretive features in connection with the development of zonal forms: solifluctional (tundra), erosional (taiga), arid—denudational (desert). However, in interpretation we must also take into account the proper (inherent) interpretive features such as joints, which characterize the internal quality of granites in the weathering zone.

Bahada surfaces found under identical late tectonic conditions of relative subsidence in tundra, taiga, and desert zones have their particular zonal interpretive features. Exogenetic processes, creating microrelief forms, the character of the vegetation, the nature of industrial or agricultural use of the region, all differ sharply in the essence and character of image pattern in all zones, forming zonal interpretive features for each one.

In addition, the totality of dynamics of the plains-forming process creates the pattern of alluvial cones, characteristic of such zones (intrinsic interpretive feature).

Geomorphic interpretation for geologic mapping is indicator interpretation. The landscape method of interpretation is used, permitting one to examine the objects of interpretation within physicogeographic complexes of different ranks.

By using complex and particular interpretive features, with consideration of morphostructural and morphosculptural features of a region, and also with consideration of the zonation of interpretive features, it is possible to make geomorphic and geologic interpretations more purposefully.

LITERATURE CITED

Bryuchanov, V. N., and Kozitskaya, M. T., "Geomorphic interpretation of Quaternary deposit," in: Aerial Methods in Geological Exploration and Prospecting for Mineral Deposits, Vol. 1, Nedra, Moscow (1964).

Chikishev, A. G., "Relation of vegetation to soil and hydrogeological conditions on Chusovaya River terraces," in: Questions of Indicator Geobotany, Izd. MOIP, Moscow (1960).

Gerasimov, I. P., Structural Features of Relief of the Earth's Surface in the SSSR and Their Origin, Izd. AN SSSR, Moscow (1959).

Gudilin, I. S., "Geomorphology," in: Geology of the SSSR, Vol. 29, Nedra, Moscow (1966).

Gudilin, I. S., and Konstantinova, G. S., "Interpretation of some complexes of unconsolidated deposits on tundra plains," in: Collected Papers on Geology and Hydrogeology (Sb. Statei po Geologii i Gidrogeologii), No. 5, Nedra, Moscow (1965).

Kobets, N. V., "Interpretation of geomorphic structure and Quaternary deposits in the region of ancient glaciation on the Russian plain," in: Complex (Integrated) Interpretation of Air Photos, Nauka, Moscow—Leningrad (1964).

Korzhuev, S. S., The Structure and History of Formation of Large Morphostructures of the Siberian Platform. Structural and Climatic Geomorphology, Nauka, Moscow (1966).

Lungersgauzen, G. F., "Features of geologic work carried out with use of aerial methods and the specific characteristics of geologic and geomorphic maps produced from this work," in: Aerial Methods in Geological Exploration and Prospecting for Mineral Deposits, Vol. 1, Nedra, Moscow (1964).

Meshcheryakov, Yu. A., Geomorphic Maps of Atlas Mira and the Principles of Their Construction. Method of Geomorphic Mapping, Nauka, Moscow (1965).

Miller, V. C., and Miller, C. F., Photogeology, McGraw-Hill, New York (1961).

Miroshnichenko, V. P., "Present-day status of the theory and practice of landscape interpretation of air photos," in: Theory and Practice of Interpreting Air Photos, Nauka, Moscow (1966).

Nevyazhskii, I. I., "Landscape science and some questions of geological interpretation," in: Landscape Science, Izd. AN SSSR, Moscow–Leningrad (1961).

Parvis, M., "Drainage patterns in identification of soils and bedrocks," Photogrammetric Engineering (1952).

Petrusevich, M. N., Aerial Methods in Geological Investigations, Gosgeoltekhizdat, Moscow (1961).

Shvyryaeva, A. M., "Geobotanical interpretation in geological investigations," in: Aerial Methods in Geological Exploration and Prospecting for Mineral Deposits, Vol. 1, Nedra, Moscow (1964).

Viktorov, S. V., "Indicator landscape science and morphometric study of landscapes," in: Aerial Surveying and Its Application, Nauka, Leningrad (1967).

Viktorov, S. V., and Vostokova, E. A., "The indicator trend in the study of landscape," in: Questions of Landscape Science, Alma-Ata (1963).

Vinogradov, B. V., "Theory of plant indicators," Byul. MOIP, Otd. Biol., Vol. 62, No. 4 (1957).

INDICATION OF ROCKS BY DRAINAGE PATTERNS
N. A. Gvozdetskii and I. P. Chalaya

Indication of rocks by air photos is an important and essential part of aerial photographic methods. At the present time it is premature to speak of an orderly and sufficiently well-developed system of methods in this respect. There are no universal, completely definite and verified interpretive features to indicate rocks. The most important premise, from the viewpoint of method, is that the indication of rocks should make use of composite interpretive features, taking into account the many-sided effects of rocks on other physicogeographic components of landscape as a whole.

However, from the sum of interpretive methods we may sort out the principal ones, which include the indication of rocks by drainage patterns (having in mind the patterns of both permanent streams and intermittent channels), a readily observed constant feature of rocks. Analysis of drainage patterns is closely related to interpretation of relief by air-photo images since it is by dissection of the rocks or hilly surfaces of the erosional network that a definite type of relief is formed. Thus, the process of rock indication by drainage patterns simultaneously includes a study of the photo image of the relief.

Factors affecting the nature of erosional dissection in a district include 1) petrographic features and mode of occurrence of the rocks, 2) bioclimatic conditions, 3) peculiarities of origin of the relief, and 4) duration of the denudation process.

Mountainous and hilly regions, according to the resistance of the rocks, are dissected to some extent by surface streams. The permeability of the surface rocks (because of jointing, porosity, etc.) is a matter of importance, diminishing surface runoff when it is high.

The steepness of sheer margins and dissection by canyons characteristic of jointed limestones subjected to karst development are related not only to the strength of the limestones and the vertical jointing, along which collapse and "renewal" of the slopes occur, but also to the permeability of the limestones, sharply diminishing surface runoff and, consequently, stream erosion of the margins. Similar features are also observed frequently in fractured sandstones and lavas.

One of the present authors has been much occupied by a study of the origin and structure of valley networks in limestones and limestone-dolomite regions modified by solution processes. Summary results of these investigations have been discussed in special papers including one by Gvosdetskii (1954, pp. 221-226). In these works note was taken of such features of valleys in limestones and limestone-dolomites modified by karst processes as their canyonlike character, their rectilinear trend and parallel sides in isolated segments, sharpness of elbow bends, sharper fractures, common persistence of several directions in the drainage and valley network corresponding to the strikes of dominant joint systems.

Such very clearly expressed features have been observed in the southern part of the Mineral'nye Vody region of the Northern Caucasus. In a figure in the monograph "Karst" (Gvozdetskii, 1954, p. 223), erosional forms are shown – gullies, ravines, rain channels – on the floors of trenches, graphically demonstrating the effect of joints, since the formation of fractures of slope retreat, being not the cause but rather the consequence of valley development, is excluded under these conditions.

Equally clear examples of the peculiarities of valley networks in limestones have been observed in the front ranges on the northern slope of the Alai Range (Akbury, Aravana, Idysai Canyons) and in the Karatau Range, and also in the Cambrian dolomites of the Angara region. Characteristic features of a valley pattern in limestones modified by karst processes may be seen on air photos of a region in Tien Shan (Fig. 1).

During investigations in the western transcaucasian region (1962), very characteristic erosional dissection of limestones was observed at the edge of the karst-modified Nakeral'skii Plateau, made up of Lower Cretaceous limestones.

Peculiarities of a valley network in limestones and dolomites, together with other interpretive features (general character of dissection, nature of steep slopes, karst forms, etc.), may represent a reliable interpretive feature even where exposures are poor and forests are common (as on the Nakeral'skii karst-modified plateau in the western transcaucasian region).

In mountains with continental climate and thin soil-plant cover, the recognition of rocks is facilitated. Investigation of interpretability of different rocks, in particular by drainage patterns, was carried out in the Tien Shan Mountains by I. P. Chalaya under the supervision of N. A. Gvozdetskii.

Interpretation of air photos of many parts of the Tien Shan has shown that different rock complexes, differing in origin and in physicochemical and petrographic properties, exhibit correspondence to specific drainage patterns and valley networks. The following groups of rocks may be distinguished: 1) Proterozoic and Paleozoic magmatic rocks, 2) Proterozoic and Paleozoic noncarbonate sedimentary and metamorphic rocks, with some inclusion of volcanic sequences, 3) Proterozoic and Paleozoic carbonate sedimentary and metamorphic rocks, 4) Mesozoic–Cenozoic noncarbonate sedimentary rocks, 5) Mesozoic–Cenozoic carbonate sedimentary rocks.

The division of rocks into five groups followed from the fact that, usually, regions composed of rocks belonging to any one of the indicated groups exhibit distinctive relief with specific forms of erosional dissection. From the drainage pattern and structural features of the relief, it is possible to determine the distribution of particular rock complexes. But, since the generalized large complexes include many rocks, it is natural that, during interpretation, we encountered a large variety of indicator features, commonly deviating from the most widespread.

We shall show some examples of interpreting rocks of the Tien Shan by drainage patterns. In the median ranges of Tien Shan, erosional, dissected, steeply-sloped landscape with a system of deep V-shaped valleys has been formed on Proterozoic and Paleozoic magmatic rocks (granitoids). The districts where the granitoidal rocks occur are distinguished by a distinctive dendritic pattern, emphasizing jointing of the intrusive massif (Fig. 2). Within the region we may distinguish isolated districts with more or less thick cover of residual and slope-wash material (patches on the photos are uniformly dark gray and gray) and exposures of bedrock, free of unconsolidated surficial deposits (white and light gray spots).

Characteristic features of rocks of the Proterozoic and Paleozoic noncarbonate sedimentary–metamorphic sequence are a palmate–dendritic drainage pattern and a generally rather dark tone on the air photos. Districts of intermediate mountains, composed of Paleozoic sand-

Fig. 1. Characteristic features of a valley network in limestones modified by karst processes. A region in Tien Shan.

stones and conglomerates, are for the most part cut by rugged valleys barren of terraces, with
a system of narrow ravines with intermittent streams. The drainage pattern on air photos is
dendritic with a common parallel–elongate form (Fig. 3, middle part of photo).

In this case, where mountainous regions are composed of systematically repeating layers
of rock, yielding differentially to weathering and erosion, a characteristic belted pattern appears on the air photos. Thus, denudation surfaces in the Boroldai Range (Karatau system of
ranges), composed of Middle and Upper Devonian sandy conglomerates and clayey rocks, exhibit a concentric banded pattern on air photos because of alternation of resistant rocks (conglomerates and sandstones) and more susceptible formations (clayey rocks). But even in such
bedded sequences, the effect of jointing is manifested in the drainage pattern.

For regions composed of Proterozoic and Paleozoic carbonate sedimentary–metamorphic
rocks, chiefly limestones, sharply dissected steep-sloped relief is characteristic, with unsystematic arrangement of rock outcrops and with a deeply incised stream network, in which the
dominant trend of joints may be observed. Limestones differ sharply from surrounding rocks
in color: they are most commonly light, in both residual accumulations and outcrops (Fig. 4).

Districts of intermediate mountains in the Alamyshik Range of the inner Tien Shan, composed of limestones and calcareous sandstones of Carboniferous and Devonian ages, are cut by
V-shaped canyons, which appear on air photos as a thin broken drainage pattern. The low moun-

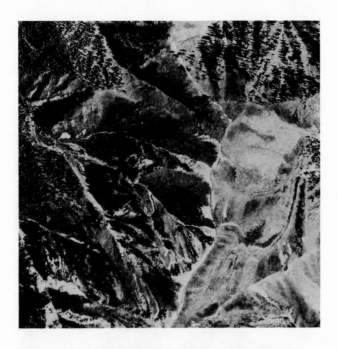

Fig. 2. Valley network in granitoidal rocks.
Intermediate mountains of the Tien Shan.

Fig. 3. Valley network in sandstone-conglom-
erate rocks. Intermediate mountains (middle
of photo) and low mountains of badlands type
(upper part of photo). Tien Shan.

Fig. 4. Valley network in limestones. Intermediate mountains of the Tien Shan.

Fig. 5. Tien Shan foothills (prilavkas).

tains of the Karatau Range in the western Tien Shan, carved in Devonian and Carboniferous limestones and dolomites are distinguished by many geomorphic and hydrographic peculiarities due to the properties of the underlying rocks.

Limestone and dolomitic low and intermediate mountains are characterized by features noted above: a) rectilinearity of individual segments of steep limestone and dolomite scarps, b) rectilinearity of individual valley segments in these rocks, usually appearing as steep-walled canyons, c) sharp elbow breaks in the plan of scarps and canyons, since the individual segments correspond to different joint systems.

The drainage pattern of regions confined chiefly to zones of faults and joints is one of the clearest and invariant interpretive features in carbonate rocks.

Mesozoic–Cenozoic noncarbonate sedimentary rocks (chiefly Neogene–Quaternary) are unconsolidated sands and fine gravels, covered by a thick layer of loessial loams, and are readily identified on air photos by the character of the relief and the drainage pattern.

The foothills of the northern Tien Shan (prilavka) and of the southwestern Tien Shan (adyr), composed of the above rocks, are characterized by gentle relief of hills and ridges with flat divides. On air photos this is reflected in an alternation of large patches and bands of different tones (Fig. 5).

The photo image of the prilavkas gives a clear impression of the gentle slopes and soft outlines of the hills and ridges. A pinnate drainage pattern is an indicator of unconsolidated Neogene–Quaternary rocks forming the lower hills (Fig. 5, upper part of picture). An indefinite dendritic pattern points to sand-granule deposits of the higher hills (Fig. 5, middle and upper part of the photo). The different type of dissection is probably associated with the thick cover of loessial loams, which increase progressively as the absolute height declines.

The inner part of the Tien Shan is well characterized by broad belts of low mountains, framing intramontane basins along the margins. Commonly a well-defined badland type of topography is found here, like that in the foothills in the surrounding Tien Shan ranges, and forms on unconsolidated siltstones and sand-granule detritus. But, in the inner Tien Shan, these rocks have other indicator features. The badlands appear on the photo as a fine banded-patchy design, consisting of a combination of narrow wavy bands of dark- and light-gray tones, corresponding to slopes having different exposures. Erosional dissection by intermittent streams has produced a dendritic pattern on the photo. The photos exhibit a fine open network, streaked with light and dark lines (Fig. 3, upper part of photo).

A separate subgroup of unconsolidated rocks according to their interpretive features consists of Quaternary deposits of various origins: sandy and gravelly alluvium, sliderock (talus, rock glacier), mudflows, moraines, and still others. The drainage pattern for these deposits may also be an indicator. For example, the eroded parts of terminal moraines commonly exhibit a braided drainage pattern with shallow but sharp erosional dissection and gentle smooth divides. Alluvial-fan deposits forming bahadas about the margins of basins are recognized by the shapes of the contours (forming fans or cones), by their occurrence at the mouths of valleys or canyons, and by the fan-like radial-braided drainage pattern of intermittent streams. An indicator of sand-gravel alluvium on the floors of intramontane basins and leveled spaces of older erosion surfaces is the wavy-linear, parallel trend and, commonly, dendritic–paniculate drainage pattern of perennial and intermittent streams.

Thus, when interpreting unconsolidated, relatively thin deposits, the accumulation of which is still going on, analysis of drainage patterns helps to distinguish such materials and to outline their distribution.

Mesozoic–Cenozoic carbonate sedimentary rocks are of limited distribution and are recognized by the same features discussed at the beginning of this article.

The above examples illustrate the dependence of drainage patterns on the rocks underlying a region. Because of this dependence, drainage patterns are important as indirect interpretive features of rocks. We may suggest that in the future more precise invariant interpretive features of different rocks will be found by thorough investigations in the field interpretation of air photos. Separation of all the large variety of rocks found in the Tien Shan into five groups may be supplemented by a more detailed classification, each step of which will be characterized by indicator features. In this classification there will be reflected many refinements of the process of identifying rocks by drainage patterns, particularly in the peculiarities of identifying particular kinds of rocks in different altitudinal zones, in different parts of a mountain system, with consideration of the specific natural conditions of the separate provinces of the mountainous region. Comparison of the interpretive features of rocks found in different mountain systems (such as the Tien Shan, Caucasus, Carpathians, etc.) also represents an interesting subject for future investigations.

LITERATURE CITED

Gvozdetskii, N. A., "The effect of jointing and karst on the development of valleys in limestones," in: Karst, Geografgiz, Moscow (1954).

THE USE OF DRAINAGE PATTERNS FOR INTERPRETING THE MORPHOSTRUCTURES OF THE MOSCOW DISTRICT

N. P. Matveev

The drainage pattern is very sensitive to different physicogeographic factors, especially to tectonic deformation and to lithology of a region. The density of drainage is a function of normal discharge, and this latter, as everyone knows, depends on a number of natural causes (Matveev, 1963, 1968).

The drainage pattern is determined in great measure by the structure of the region. In the Moscow region, the upper course of the Nara (from the mouth of the Ist'ya to Proletarskii) and the upper reaches of the Lopasnya, Kashirka, Mocha, and Rozhai have southeastern trends, corresponding to the trend of the Moskvoretsko-Oka ridge. This fact cannot be considered accidental, since this pattern could only arise through the effect of a single morphostructure on the stream courses.

Tectonics and lithology affect not only the drainage pattern but also the longitudinal profile of the streams. In the present paper the author uses the relations in the Moscow region to examine the relationship between longitudinal profiles of streams and structure and to consider the possible interpretation of morphostructures from such longitudinal profiles.

There is no single opinion at the present time concerning the effects of tectonics and lithology on the longitudinal profile of a stream. Popov (1965) has stated that recent tectonic deformation has been so slight in comparison with erosion that it could scarcely have any serious effect on altering the profile. We can agree with the opinion of Popov only in part. When the stream profile is youthful, when the gradient is large, erosion is also great. Recent tectonic deformation embracing a small area apparently does not substantially affect the stream profile. But a completely different picture is observed when the profile is near equilibrium. Downcutting is absent in this case, and tectonic deformation may change the profile. It cannot be otherwise. If the opposite were true, the stream profile, having reached the equilibrium state, would remain now and always in the same state, and streams as a whole would bear the imprint of senility. In nature such systems do not exist.

More than this, we cannot deny the effect of tectonics also on young stream profiles. Streams usually cut across regions differing in their structural relations, and the longitudinal profile must therefore reflect both positive and negative structures. If these structures are reactivated, erosion is hardly able to cut down completely and to produce a profile corresponding to the hydraulic conditions of the stream.

With tectonic deformation, as is well known, the gradient of the longitudinal profile changes. The change, however, is not the same everywhere along the profile. Downstream the gradient increases, upstream it declines. Consequently, erosion increases only where the gradient increases and diminishes where the gradient declines. Beyond this, all will depend on the magnitude of the deformation, and also on the rocks that occur in the stream bed. If deformation involves a great part of the profile, and if resistant rocks occur here, the stream will not be able to cut down quickly through the deformed segment.

On the whole, the lithology determines the nature of the longitudinal profile. In details, the lithology has little effect, however, except in the upper reaches of the stream. Breaks in the longitudinal profile of a stream are due to more resistant rocks. Weak rocks play a subordinate role. The amount of downcutting is determined by resistant rocks. It may be shown that streams exhibit selective erosion, i.e., the stream cuts deeper in weak rocks that are more susceptible to erosion, which should lead to the appearance of troughs and waves on the profile and to the formation of irregular gradients. The gradient should be increased in the resistant rocks, diminished in the weak rocks, and, consequently, erosion of the benches should occur and accumulation begin in the troughs.

Ultimately the profile will assume a smooth curve. In this case we have a notable example of self-regulation of the system: the stream of water affects the profile, and the profile affects the rate of flow. As we know, the stream velocity is a function of gradient; with increase in gradient the kinetic energy increases; with decrease it declines.

Diminution of gradient in zones of weak rocks stops erosion; increase of gradient in resistant rocks increases erosion and diminishes the gradient. Concerning the statement that lithology has no effect on the selective capacity for erosion, this is attested by the fact that the Protva River cuts through rocks differing in resistance: clays, sands, and limestones. In all, there are five horizons represented by limestones, four of clays and sands. On the basis of general views, that resistant rocks form ledges, weak rocks, low areas, there should then be found five ledges and four low areas between Borisovo to the mouth of the Protva. However, the longitudinal profile of the Protva exhibits but one ledge, at Spas–Zagor'e below the falls of the Luzha. Thus, even such variation in lithology as present here proves to have no great influence on the shape of the Protva profile.

Erosion of weak rocks cannot exceed some limit, determined by the amount of moving water. Let us assume that a stream has eroded weak rocks to a depth of 5-6 m as compared with a bed on resistant rocks. The length of the eroded segment will be taken as 1 km, the width of the stream as 100 m, and the average depth as 4 m. The volume of the channel cut amounts to 400,000 m³. Let us assume that the stream velocity in the resistant rocks is 1 m/sec. Discharge of water amounts to 400 m³/sec. We find the velocity of the mass of water in the excavated channel to be

$$\frac{400}{g} \times 1 = \frac{400,000}{g} \times v,$$

whence

$$v = \frac{400}{400,000} = \frac{1}{1000} \text{ m/sec.}$$

At a velocity as low as this, erosion would be impossible.

If we start with the relation between stream velocity and diameter of particles established by Velikanov (1958), we may find the limit of how much weak rock can be eroded. Erosion stops at a stream velocity of 26 cm/sec. Actually, erosion virtually stops at even greater velocities. For example, the erosive velocity of water for sand having a diameter of 1 mm is 46 cm/sec. Self-regulation manifests itself not only in development of the longitudinal profile of a stream but also in the formation of meanders. The growth of meanders leads to their destruction and to straightening of the stream.

In principle, tectonic disturbance of the longitudinal profile also causes self-regulation, but this process, in contrast to the effect of lithology, is not passive. It leads, on the contrary, to deformation of the profile. The rate of deformation and also the velocity of the stream are determined by the dimension and magnitude of the structure. If the rate of uplift of the structure exceeds erosion, the structural effects continue to grow.

Besides tectonics and lithology, physicogeographic factors also affect the longitudinal profile of a stream: temperature, precipitation, evaporation, snow cover, moisture deficit, forest growth. These affect the amount of runoff and thus determine the nature of the longitudinal profile. The amount of normal runoff determines the general character of the longitudinal profile of a stream. The shape of the profile depends on discharge of the stream, and the discharge of the stream with the influx of tributaries increases in jumps, and because of this the profile of the stream exhibits breaks. We should not forget, however, that many streams cut valleys along structural lines. The tectonic effect is therefore commonly accentuated by the hydrological effect.

Several longitudinal profiles of a stream may be considered: a) the actual profile of the stream, b) the generated profile, corresponding to the hydraulic conditions of the stream, c) the profile of equilibrium corresponding to the present-day base level of the stream and present-day runoff, d) the profile of equilibrium corresponding to present-day runoff and a single base level for the given river system.

The number of longitudinal profiles of equilibrium of a stream may be multiplied endlessly, each corresponding to a particular stream discharge. At flood time, the stream tends to develop a profile corresponding to greatest stream discharge. The stream erodes strongly in this case. At low water the stream tends to create a longitudinal profile corresponding to the given discharge and the given hydraulic regimen of the stream. In considering that the discharge changes during the year and experiences fluctuations from year to year, such profiles theoretically exist in great numbers, but all of them lie between two extremes, corresponding to the minimum of annual least discharge and maximum of annual greatest discharge. As is known, the recurrence of discharge is not uniform: some discharge rates repeat frequently, others are very rare.

The highest recurrence is characteristic of discharge near the average over a period of many years. The greatest probability, therefore, is that the profile of equilibrium will lie nearest the profile corresponding to the average of a period of several years. However, according to the fluctuations in runoff, the stream will not tend continuously toward a profile of equilibrium, but, in approaching or diverging from the equilibrium profile, will still tend steadily toward it. Even when the stream reaches a profile of equilibrium very frequent deviations occur because of fluctuations in runoff, discharge, and general water content in the streams. Longitudinal profiles of intermittent streams lie below the equilibrium profile, at times above it, at times coincident with it. Thus, a stream never ceases eroding and accumulating, even in the equilibrium state. It is thus more nearly correct to have in mind, by longitudinal profile of equilibrium, not a state of static equilibrium, but a state of dynamic equilibrium, when erosion gives way to accumulation, but these processes fluctuate about the state of equilibrium. We have found an equation that describes the longitudinal profile of a stream (Matveev, 1968).

$$H = H_0 - A_{m_0^2} \ln x,$$

where H is the ordinate of the stream profile (height); H_0 is the elevation at the headwaters of the stream; A is a parameter characterizing the physicogeographic aspects of the stream; m_0 is the normal runoff at the point computations of the stream profile are begun; and x is the length of stream from the headwaters to the point of computation. The technique of computing parameter A has been discussed in our earlier works (Matveev, 1967, 1968).

In comparing the first three types of stream profiles, we may note the deviations of the actual profile from the theoretical (generated). The deviations may be either positive or negative. In plotting these deviations on a map and comparing them with the geologic structure of the region, we may mark morphostructures and zones of recent tectonic movement.

We call the negative zones negative anomalies, the positive zones positive anomalies. The arrangement of positive and negative anomalies is not unsystematic, but rather follows something of a pattern. Negative anomalies occur chiefly in the upper Volga lowland and the Meshchera, i.e., the zone of very low relief. Isolated segments of negative anomalies are observed between Ruzaya and Staronikolaev in the upper reaches of the Protva and Lusyanka, between Cherlenkov and Zhelomein on the Ruza. A large zone of negative anomalies is observed also along the middle course of the Protva, Nara, and Lopasnya Rivers, between Naro-Fominsk and Naro-Osanov, in the middle course of the Moskva (Moscow) River (from Andreevskii to the Istra), along the middle course of the Pekhorka, on the Klyaz'ma, between Pikrov and Losino-Petrovskii, and along the lower reaches of the Malaya Istra. Apparently all the negative anomalies should be assigned to zones of subsidence. Positive anomalies are grouped in more distinct zones than the negative anomalies. These zones of positive anomalies have an arcuate distribution, corresponding to the trend of the limbs of the Moscow basin (syneclise).

A belt of positive anomalies passes along a line trending almost north to the west of a zone through Ul'yanovo on the Shosha, Volokolamsk, Spas, Ostashevo on the Ruza, Porech'e on the Moskva, Mozhaisk, and Shustikovo in the Vereya region. From this north-trending belt to the east along the Moskva to the Istra, the zone of positive anomalies trends eastward. A second belt of positive anomalies passes from the headwaters of the Dubna to Zagorsk, Khot'kovo, Lobnya, Zelenograd, Istra, Zvenigorod, Aprelevka, Klimovsk, Molodi, Rastunovo, Nikonovskoe, and Meshcherino.

Isolated positive anomalies are observed south of Obninsk and Maloyaroslavets, in the headwaters of the Protva and Koloch' Rivers, and at Bol'shie Gorok and Turginovo on the Shosha. The positive anomalies correspond clearly to zones of uplift. Apparently the zones of positive and negative anomalies correspond to a broad tectonic plan on which uplift and subsidence of smaller scale are superimposed. For example, zones of subsidence are characterized by small segments of uplift, and zones of uplift are characterized by small segments of subsidence.

We have grouped the zones of uplift and subsidence into individual zones regardless of where they occur, in the zone of negative anomalies or the zone of positive anomalies. The overall result appears almost the same, but the entire scheme of runoff becomes much more orderly. The zones of subsidence also assume a semicircular form. A large zone of subsidence extends from Gubino through Ruza, Tuchkovo, Nara-Osanovo, Kubinka, Lenino, Novo-Mikhailovskoe, Voronkovo, and Khodaevo. Within the Klin-Dmitrov Ridge, zones of subsidence along an east-trending line have been observed. One of these passes through Pyatnits, Belavino, and Glazovo, another through Krasnoarmeisk, Pushkino, Mytishchi, and Kaliningrad, clearly separating the Shchelkovo–Fryazino uplift from the large zone of the Zagorsk–Odintsovo–Klimovsk uplifts.

LITERATURE CITED

Matveev, N. P., "Formation of drainage patterns in the southern part of the Moscow Oblast," in: Soil Science, Vol. 6 (XILVI), Izd. MGU (1963).
Matveev, N. P., "Use of quantitative methods for investigating longitudinal profiles of rivers in the Moscow region," Byull. MOIP, Otd. Geol., Vol. 62, No. 4 (1967).

Matveev, N. P., "Geomorphic analysis of longitudinal stream profiles in the Moscow Oblast,"
    Uchenye Zap. Mosk. Obl. Ped. Inst., Obshchaya Fiz. Geografiya, Vol. 181, No. 12 (1968).
Popov, I. V., Deformation of Stream Beds and the Construction of Hydroelectric Power Plants,
    Gidrometeoizdat, Leningrad (1965).
Velikanov, M. A., The Channel Process, Fizmatgiz Lit-ry, Moscow (1958).

SOME NEW IDEAS AND CONCEPTS IN THE STUDY
OF INDICATORS
D. D. Vyshivkin

For a number of years, a course called Science of Indicators has been given in the Section of Biogeography of the Department of Geography at Moscow State University (MGU). The course is based on the ideas of indicator geobotany rather thoroughly discussed in the book by Viktorova, Vostokova, and Vyshivkin (1962). In recent years, however, these views have undergone some modifications, and some of the concepts and determinations examined in the indicated work have been refined.

Concept of the Object of Study in Indicator Investigations. During recent years, the concept of the basic object of indicator investigations has become rather clearly crystallized: the system of indicator–indicated object. This view is reflected in the discussions of the 3rd Conference on Photo-Indicator Research (Leningrad, January 1967), and also in the dissertation of Vostokova (1967).

Classification of Indicated Objects. In works devoted to the theory of indicator investigations, considerable attention is generally given to the classification of indicators (Viktorov, 1955; Vinogradov, 1964), but, up to the present, there has been no classification made of indicated objects from the point of view of indicator relations. Usually they are distinguished on the basis of which components of the geographic environment or which properties of the components it is necessary to determine. However, the relationship between indicator and indicated object may be very complex and variable. In some cases the indicated object is an important ecological factor, proving to have a great influence on the plant cover or playing a leading role in landscape processes. At other times, the indicated object, although of great significance in the landscape, acts on the indicator not by one particular factor but by a combination of factors. And, lastly, it is also possible that the indicated object is not directly related to the indicator and that it may be manifested only by features accompanying it. In this respect, all indicated objects may be divided into two groups: monofactorial and polyfactorial.

Monofactorial objects may include indicated objects having a strong ecological effect on vegetation or factors playing a very significant role in landscape development. Most commonly these will be either individual elements and specific substances (such as bitumens) or individual properties of natural bodies (such as mineralization of groundwater, salinization of soil, and the like). Commonly, indicator features are complicated by the effect of a monofactorial indicated object (such as the development of morphological deviations in plants because of the influence of bitumens).

Many indicated objects are not leading ecological factors and the indication relationship between indicated object and indicator is realized by means of the summed ecological effect of a number of factors. Such objects may be called polyfactorial. By their nature polyfactorial

118

indicated objects are also inhomogeneous and are represented by two groups: 1) integral poly-factorial indicated objects, and 2) autonomous polyfactorial indicated objects. Integral objects include those in which different factors interact so closely with each other that only their integrated effect determines the general effect of the given object, defining its specific properties. In this case any individually selected factor may be changed only within definite limits. With appreciable change of any particular factor, the general properties of the indicated natural body are changed, and the body shifts to a different taxonomic category. An example of an indicated object of this type is soil, the ecological effect of which is determined by the entire sum of its properties: mechanical state, content of nutrients, salts, distribution of these components in the profile, and others. With considerable change in any of these properties, the given soil variety changes to a different variety.

Autonomous polyfactorial indicated objects include objects that are indicated only by a series of indirect features. The objects do not depend on factors by which they may be mani-fested in a given region. In this type of indicated object, the most extensive change in any fac-tor leads to no change in the indicated object. An example of such an object is the indication of geologic age of a rock. For each determination in a region, indication of rocks of a definite age is possible in connection with the fact of definite lithology, chemistry, thickness, development, distinguishing them from rocks of a different age. However, rocks of any particular age may undergo facies changes, producing a new set of indicators.

It should be noted that the separation of objects into monofactorial and polyfactorial re-flects their common property and not the actual method of revealing them, since in actual cases it is possible that it will be advisable to use indirect indicators to reveal a monofactorial indi-cated object. For example, we may recognize a region of fresh-water occurrence by the distri-bution of weakly indurated sands with single psammophytes. At the same time, the distribution of some tracts and facies are closely associated with definite polyfactorial indicated objects. The relationship may be so close that the origin of the given indicated object is connected with definite landscape conditions, including physiognomic components of the landscape. Thus, the origin of different soils, along with the lithogenic basis and climate, is a function of the plant cover and the relief of the locality.

Evaluation of Physiognomic Quality of Indicators. The concept of "physiognomic quality of indicators" (physiognomicity), introduced by Viktorov, Vostokova, and Vyshivkin (1962), is an important characteristic of indicators, since, along with readily discernible indicators, there may be great practical value in poorly discernible indicators and indicator features, which may possess high reliability and wide distribution in the investigated region. That is, they may have high significance. Such indicators and indicator features in-clude morphological anomalies of vegetation, for the reliable determination of which it is some-times necessary to resort to mass measurements, individual rare species, and communities having similar aspect but different indicator significance. The aspects may be alike only in certain seasons, such as communities of red feather grass (Stipa rubens) and desert oat (Avenastrum desertorum), associated in northern Kazakhstan with different types of habitat, readily discernible in the first half of the summer but having identical aspects after the plants have gone to seed. Recently we have begun to use poorly discernible morphological features of the relief for indicator purposes: peculiarities of the finest microrelief (nanorelief), nature of jointing, trend and density of drainage pattern, and so forth.

In connection with the fact that the physiognomic quality is difficult to evaluate by quanti-tative indices, we may propose only the most general qualitative evaluation, taking into account how the indicator may be recognized. 1. High physiognomic quality; the indicator may be rec-ognized by aerial methods or by surface traverses in a car or truck. 2. Moderate physiognomic quality; indicator may be recognized only on surface traverses (by foot). 3. Poor physiognomic quality; indicator may be recognized only by investigation at isolated points, involving measurements and calculations, and comparison of such data with those obtained at other points. 4. Very poor physiognomic quality; for recognition of the indicator it is necessary to set up special experiments or to conduct prolonged observations (such as determining microclimatic conditions by the state of seedlings sown in special growing vessels, as in the experiments of Il'inskii (1939)).

Value of Individual Characteristics of Indicators (reliability, significance, physiognomic quality). The principal index, determining the possible use of any natural component as an indicator of a definite indicated object, is reliability, which determines both the possibility of using the given component as an indicator and the precision of the prognosis.

Significance is necessary for selecting indicators to communicate with other specialists; i.e., it must be considered in preparing various reference herbariums, photographic collections, and other materials, because inclusion of indicators that are not widespread in such reference works may only complicate the use of such works by specialists.

The physiognomic quality of an indicator has a double meaning. On one hand, it is necessary to keep in mind when preparing reference books in which it is advisable to include only indicators having high and moderate physiognomic quality. On the other hand, the physiognomic quality determines the site of indicator investigations in the general work and it determines the size of area in which it is possible to carry on the indicator investigations.

In using indicators having high physiognomic quality, indicator investigations may be preceded by other types of investigations or may be carried on simultaneously with them over very large areas. In this case, preliminary separation is made of regions where detection of the indicated object is most likely but not necessary for indicator observations. Investigations by means of indicators having moderate and, especially, poor physiognomic quality are best made in regions where detection of the indicated object is most probable. These regions are discovered by other methods, and have frequently been pointed out already by specialists making investigations. Indicator investigations in this case should refine the suggested prognosis. Thus, to carry out botanical indicator investigations on the discovery of zones with high bituminous content, effected by discovery of morphological deviations of plants, it is advisable to work in areas which, on the basis of general geological exploration, have been recognized as promising for oil and gas and need only the collection of supplementary special information. Regions in which investigations have been made by means of indicators having poor physiognomic quality should be very restricted in size.

Key Areas in Landscape Indicator Investigation. Considerable attention in the course at Moscow State University has been given to landscape indicator investigations, which we understand to mean the use of physiognomic components of the landscape for recognition of decipient components (Viktorov and Vostokova, 1963). Landscape-indicator investigations not only refine and facilitate the discovery and evaluation of indicated objects, but they have also fundamental distinctive characteristics. Thus, in making geobotanical-indication investigations we may evaluate features of the indicated object only in areas where the indicator occurs.

For landscape-indicator investigations using a combination of direct and indirect indicators, it is possible to evaluate distribution characteristics of the indicated object, subject to migration, within the entire landscape and to trace both the path of migration and the change undergone by the object during the migration. These questions were studied in special detail by Vostokova (1967) in relation to groundwater. In regard to the migration of individual elements, these statements find their reflection in the science of the geochemistry of landscapes.

In connection with the manifestation of landscape-indicator investigations, at the core of methods for indicator investigation – the method of standards and the method of ecological profiling – we should add the method of key areas. A key area is defined as the principal facies of a tract or landscape of which an indicator investigation is being made, the mutual interrelations of these facies, and also the role they play in the migration of the indicated object. Thus, in setting a key area for indication of water in a region of gullies and ravines, one must study the features of the principal channel of a ravine, beginning from a point above where water first appears to a point where some aquifer discharges, must study lateral hollows or gullies of the ravine, both those showing evidence of carrying water and those showing no such features, and must plot the profile from the divide between ravines to the thalweg of the ravine. In describing the features one uses the method of ecological profiling, and also standard or detailed-traverse descriptions. All these profiles and descriptions are combined to characterize a single area or plot of the landscape.

It should be noted that the wide development of landscape-indicator investigations does not exclude the development of other types of indicator investigation (botanical, geobotanical, geomorphic, and others), since there may be objects that are rather clearly and simply revealed by individual physiognomic components of the landscape, not requiring any composite study. Furthermore, the recognition and description of a series of indicated objects by landscape methods are generally impossible. For example, determination of the age of a landslide is made by means of annual rings on trees, i.e., by a purely botanical method (Turmanina, 1964). Thus, we should not speak of the replacement of geobotanical-indicator investigations by landscape methods, but rather of the development of landscape-indicator methods along with further refinement of particular indicator investigations.

At the present time, it is clearly necessary to produce a general textbook on the entire subject of the science of indicators. In addition to the questions briefly discussed above, such a textbook should be devoted also to the many new methods used in the different kinds of indicator investigations.

LITERATURE CITED

Il'inskii, A. P., "The use of plant indicators in studying bioclimates," Izv. Gos. Geogr. Obshch., Vol. 71, No. 5 (1939).

Turmanina, V. I., The Interaction of Vegetation with Landslide Processes on Slopes as Exemplified in the Landslides of Moscow and Its Environs, Author's abstract of candidate's dissertation, Moscow (1964).

Viktorov, S. V., The Use of the Geobotanical Method in Geological and Hydrogeological Investigations, Izd. AN SSSR, Moscow (1955).

Viktorov, S. V., and Vostokova, E. A., "Indicator trend in the study of landscapes," in: Questions of Landscape Science, Alma-Ata (1963).

Viktorov, S. V., Vostokova, E. A., and Vyshivkin, D. D., An Introduction to Indicator Geobotany, Izd. Moskovsk. Gos. Univ. (1962).

Vinogradov, B. V., Plant Indicators and Their Use in Studying Natural Resources, Vysshaya Shkola, Moscow (1964).

Vostokova, E. A., Theoretical Basis of Landscape- and Water-Indicator Investigations and the Technique of Using Them in Prospecting for Groundwater in Deserts, Author's abstract of doctoral dissertation, Moskovsk. Gos. Univ. (1967).

MICROPHYTOCENOSES OF SOME LANDSCAPES
OF NORTHERN SIBERIA AND
THEIR INDICATOR SIGNIFICANCE

N. G. Moskalenko

A characteristic feature of the plant cover of northern Siberia is its mosaic character. This feature in the structure of northern plant communities was studied specially by Norin (1962, 1965), Petrovskii (1962), and Minyaev (1963). These investigators and geobotanists, having studied the mosaic structure of forest communities of the temperate zone, steppes, and deserts and the phytocenoses of mountainous tracts (Yaroshenko, 1934; Sakharov, 1950; Lavrenko, 1952; Levina, 1958), showed that the presence of microgroups (microphytocenoses) responsible for the mosaic structure of the plant cover are usually associated with inhomogeneity of environmental conditions. Another cause leading to mosaic structure in vegetation is the growth of individual species of plants, forming plots.

Since the basic cause leading to plant communities of microphytocenoses is inhomogeneity of the surrounding environment, it should be expected that the study of microphytocenoses in indicator investigations permit us to refine the indicator significance of geobotanical and landscape indicators. Such refinement may be of interest for large-scale engineering-geological and geocryological mapping. Investigation of microgroups is also interesting because it permits us to evaluate past and possible future plant communities. Thus, even in little-used regions of Siberia we must contend with a large number of secondary plant associations (such as in burned-over areas of various ages). A study of the components of the microassociations in these areas permits us to determine to which of the native associations they should be assigned and what natural conditions are characteristic of them.

During indicator investigations by the All-Union Scientific-Research Institute of Hydrogeology and Engineering Geology (VSEGINGEO) in the northern part of Western and Central Siberia, we repeatedly noted mosaic structure of the plant cover in connection with soil, frozen ground, and other environmental conditions. The investigations were conducted by the technique usually used for indicator studies (Viktorov, Vostokova, and Vyshivkin, 1962). For study of the mosaic structure, determinations were made specially of the frequency of species in areas of 0.1 m^2 along a linear traverse, but sometimes horizontal projections of the plant cover were made for meter-sized areas.

In Western Siberia our work was conducted in the Pur and Novyi Port regions of the Tyumen' Oblast. In the Pur region investigations were conducted along the Pur valley in the segment from Tarko-Sale to Samburg, in the subzones of northern taiga and wooded steppe. Six key areas, selected by the method worked out by Mel'nikov and his coworkers (1966), were studied here in 1967 jointly with the geologists A. N. Kozlov and V. A. Levandovskii. Later, on the basis of data obtained from surface field work in these areas, aerial reconnaissance, and interpretation of air photos, a schematic landscape-indicator map of the left bank of the Pur was prepared. A part of this map, for one of the key areas, is shown in Fig. 1.

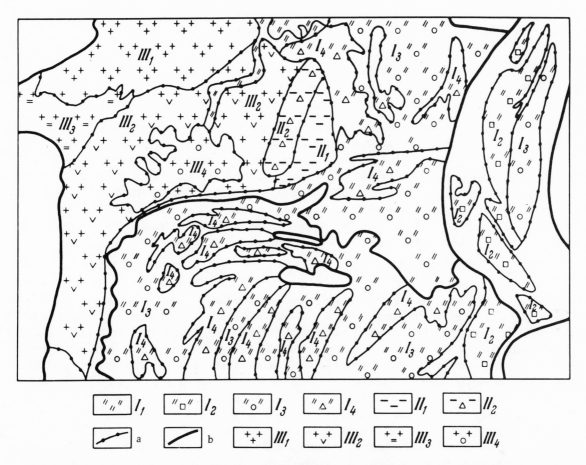

Fig. 1. Part of a schematic landscape-indicator map of the area along the left bank of the Pur River. I_1) Flood plains of small streams with shrubs, meadows, and sedge swamps; sand and sandy loam, thawed; I_2) low flood plain of the Pur, locally covered by willow groves and meadows; very-fine-grained and fine-grained sand, thawed; I_3) high flood plain of the Pur, overgrown with birch–Siberian pine and Siberian pine–birch–spruce forest; very-fine-grained and fine-grained sands with layers of sandy loam and loam, thawed; I_4) meander-belt lowland of flood plain with sedge and birch–sedge swamps; loam underlain by very-fine-grained sand at a depth of 0.5-1 m, peat; II_1) smooth, weakly dissected surface of first terrace above the flood plain, with larch–pine and birch–pine–Siberian pine thin forests; very-fine-grained and fine-grained sand, thawed; II_2) flat swampy surface of 1st terrace with flat–hummocky peat bogs and ridge-and-seep swamps, peat; III_1) slightly hilly surface of 2nd terrace, with birch–Siberian pine–larch thin scrubby forest and small segments of tussocky scrub–cloudberry–sphagnum peat bogs in low places; loam underlaim by fine-grained sand; III_2) slightly hilly surface of 2nd terrace with thin birch–Siberian pine–larch forests near edge of terrace; very-fine-grained sand with layers of loam; loam underlain by fine-grained sand; III_3) flat swampy segments of 2nd terrace with complex lichen–shrub peat bogs, peat; III_4) flat swampy segments of 2nd terrace with sphagnum–lichen and cloudberry–Ledum flat-tussocky peat bogs; peat; a) landscape boundaries; b) streams, basins.

The valley of the Pur, the geologic structure and geomorphology of which are described in the paper of Andreev and Stal' (1960), in the investigated segment, has a well-developed flood plain up to 4-6 km wide, covered by willow groves mixed with forests of pine, birch, spruce, and larch, sedge meadows and swamps.

The Pur flood plain adjacent to the channel is characterized by horsetail and reed—grass willow groves, developed on flood-plain bedded alluvial soils. The soil-forming materials here are very-fine-grained and fine-grained sands, and groundwater occurs at a depth of 1.5-3 m.

Sedge swamps on peaty and muddy swamp soils are widespread on the flat smooth surface of the central flood plain and at sites of meander scars. In August the groundwater under this vegetation is found at a depth of 0.8-1 m. On better-drained segments of the central flood plain, small scattered areas reed-grass and sedge meadows on flood-plain alluvial—meadow soils. The groundwater in such areas was found at a depth of 2-2.5 m. Small flood-plain peatbogs found on the flood plain are covered by birch—sedge communities, under which groundwater is found at 0.3-0.5 m at the end of summer. Soil-forming material beneath the flood-plain meadows and swamps is loam, underlain at shallow depth (0.6-2 m) by sand.

The crests of the flood plain, having a height up to 4 m and a width of 10-30 (up to 100) m, are covered by various associations of mixed forests. Mixed forests with lingenberry—cowberry—lichen patches and bilberry—cowberry patches are found on flood-plain, podzol—leached-ferruginous soils on very-fine-grained and fine-grained sands. Groundwater occurs at a depth of 2-4 m under these forests. Mixed forests with lingenberry—raspberry—reed grass patches, on flood-plain bedded alluvial clayey soils, occupy the largest areas of the flood-plain forests. The soil-forming material in these areas is very-fine-grained and fine-grained sands with layers of loam and sandy loam. Groundwater has been found to lie at 3-4 m. On similar soils in the northern taiga, one commonly finds zones of birch—larch and birch forests with some Siberian pine and larch forests with sedge grass—raspberry—lingenberry patches. Beneath these forests permafrost may be found at a depth of 2-2.2 m at the end of August. Within the forest-tundra zone, these flood-plain forests are rather widespread. Spruce—birch and larch—spruce forests with lingenberry and cowberry—haircap moss patches occupy small areas of peat—podzol soils on thin sandy loam and loam, underlain by frozen sands with layers of sandy loam and loam. The depth of seasonal thaw beneath these forests is 0.8-0.9 m.

Hummocky frozen peatbogs are widespread along the terraces of the Pur River. The hummocks of these bogs contain sphagnum—lichen cloudberry—Ledum communities. The depth of seasonal thaw beneath these communities amounts to 0.35-0.7 m. Sedge—sphagnum communities occupy the thawed swampy zones between hummocks. The thickness of peat in the hummocks reaches 1.5-3 m.

Thin forests and sparse stands of larch, birch, Siberian and other pine are found on drained, slightly hilly areas of the terraces. Larch—pine, larch—Siberian pine, and birch—larch Ledum—lingenberry cowberry—lichen thin forests grow on podzol—leached-ferruginous soils on very-fine-grained and fine-grained sands. Beneath these forests the groundwater is encountered at a depth of 7-8 m. Peat-podzol soils on loams underlain by fine-grained sands are sparsely covered by birch—cedar (Siberian pine)—larch, horsetail—Ledum—lingenberry, cowberry stands. Groundwater under these phytocenoses has been found at a depth of 7-8 m.

Larch—Siberian pine, dwarf birch—lingenberry, haircap moss—lichen sparse forests and thin woods and also larch—birch and larch, Ledum—lingenberry, cowberry—lichen thin woods occur on segments of the terrace that are more poorly drained, on gley—podzol soils. Soil-forming materials in these areas are very-fine-grained and fine-grained sands with layers of sandy loam and loam. The water table lies at a depth of 1.5-3 m.

In the Pur region, a clear example of mosaic structure of the northern phytocenoses is the horizontal distribution of thin larch—cedar (Siberian pine) and larch—pine, bearberry—Ledum—lingenberry forests with moss—lichen cover, developed near terrace edges and on the relatively high terrace surface with hill-and-swale relief. The water table here lies at a depth of 7-8 m, and the soil-forming material is sand. Woody species in these thin forests are cedar (Siberian pine), other pine, and larch, with some birch. The closeness of the crowns of woody genera amounts to 0.1-0.2 m; the height is 8-12 m.

A clear mosaic structure is expressed in the surface and soil cover of the thin forests. The structure is due to the presence usually of three microgroups associated with different elements of the microrelief. On hillocks 1-2 m high and 3 × 6 m in plan view, we found the Ledum—lingenberry—bearberry—cladonia microgroup, beneath which occurs a cryptopodzol soil. The cover of shrubs in this microgroup amounts to 30-40%, the cover of lichens 100%. The lichen cover contained chiefly Cladonia alpestris (L.) Rab., but Cl. rangiferina (L.) Web., Cl. coccifera Wild., Cetraria cucullata Ach., and others were also present. Basins, as deep as 2 m and horizontal dimensions of 3 × 4 m, and the lower slopes of the hillocks are the sites of Pleurocium—crowberry—bilberry, Pleurocium—crowberry—bilberry—Siberian pine, or pine microgroups. Beneath these microgroups, under conditions of more intense surface moisture, medium podzol leached-ferruginous soils were formed. The shrub cover in these microgroups amounts to 60-70%, moss to 90%. In the soil cover Pleurocium Schreberi is the dominant form; Dicranum sp. and Polytrichum sp. are characteristic among other mosses. In these phytocenoses of basins, shrub vegetation is better developed than on the hillocks. For example, bilberries and lingenberries are 5-7 cm high and more on the average.

Thin larch—Siberian pine, dwarf birch—lingenberry woods with some birch and a moss—lichen cover occur on the surface of a terrace with a more or less well-expressed hill-and-swale relief and with shallow groundwater (1.5-3 m), confined to sands with layers of loam and sandy loam. These plant communities differ from those described above chiefly by greater spacing of trees and by the composition of the soil cover, in which polar birch plays an important role. Ledum—lingenberry—bearberry microgroups on cryptopodzol soils are developed on hillocks 0.5-1 m high, some as high as 2 m, up to 10-15 m across, as in the thin forests described above. In some places Ledum is not characteristic of the microgroups of the hillocks. The basins contain Polytrichum—dwarf birch microphytocenoses on gley—podzol soils. Moss cover in these microphytocenoses consists of Polytrichum commune L., P. strictum Banks, Dicranum sp., and several other mosses. The shrub cover is dominated by polar birch, various bilberries (Vaccinium ulaginosum, V. myrtillosum), and bulbous sedge.

On black and white panchromatic air photos, the described differentiation of the plant cover into microphytocenoses, restricted to different microrelief elements, is represented by a characteristic reticulated or cellular pattern. The tone of the photo image of microgroups on hillocks is gray and light gray; microgroups in swales appear medium to dark gray.

Microgroups in complex surface cover of cloudberry—lingenberry—sphagnum—lichen flat-hillocky peat bogs are also clearly distinguished. These peat bogs are found in the Pur region on flat, swampy terrace surfaces. They are frozen, the temperature of the permafrost zone being 0.2-4°.

Three microgroups are developed for the most part on hillocks of cloudberry—Ledum—sphagnum—lichen peat bogs. Hillocks formed of mosses and having a height of 0.2-0.5 (up to 0.7) m and a diameter of 0.5-1 (up to 2-3) m bear the sphagnum—Ledum microassociation. For this microassociation a continuous cover of sphagnum mosses is characteristic, commonly with Dicranum and Pleurocium. Shrubs are well developed, covering up to 90% and consisting chiefly of Ledum. The depth of thawing beneath this microgroup is 0.4-0.6 m. The areas between the hillocks are covered by the andromeda—lingenberry—cladonia microgroup. This microgroup is distinguished by solid development of a lichen cover consisting of different species of Cladonia: Cl. alpestris (L.) Rab., Cl. rangiferina (L.) Web., and others. Cetraria islandica Ach. and some other lichens are also present. Shrubs here are scarce, the cover of shrubs usually representing no more than 30-40%. Besides andromeda and lingenberry, the shrub cover commonly contains bog bilberry, cranberry, featherleaf (Chamaedaphne), and other forms. The shallowest depth of seasonal thawing, 0.35-0.45 m, is found beneath the andromeda—lingenberry—cladonia microgroup, and the soil temperature is lowest. Measurements on soil temperatures at depths of 5-15 cm beneath this microgroup gave values of 3-4.5° and lower than at the same depths beneath the sphagnum—Ledum microgroup.

Cloudberry—sphagnum or dwarf birch—cloudberry—sphagnum microgroups occur in shallow channels along frost cracks and small seepage areas. The latter microgroups were also encountered on the slopes of the hillocks. The sphagnum mosses in these microgroups forms a continuous cover, in which appear scattered cloudberries and some other shrubs, forming no more than 40% of the cover. The depth of thawing beneath the cloudberry—sphagnum microgroup is 0.5-0.7 m. The differences in depth of seasonal thawing and in temperature of the soils beneath the described microgroups are due both to the effect of very fine microrelief, formed either by the growth of sphagnum mosses (hillocks) or by frost cracking and thawing of frozen ground (grooves, swales), and by peculiarities of the plant cover itself, spatially inhomogeneous in its reflectivity, amount of cover, saturation with water, and some other indices.

The horizontal segregation of phytocenoses into microgroups was studied by the author in the tundras of the Novyi Port region of the Tyumen' Oblast. Investigations in this region were carried out in 1966 in the vicinity of Novyi Port, embracing the treeless tundra along the Gulf of Ob and the first three terraces. The natural features of the region investigated have been described briefly by Ivanova (1962), Orlova (1962), Zhukov and Saltykov (1953). The landscape units we distinguished in the Novyi Port region and their indicator significance are shown in Table 1.

Lichen—moss cotton grass—dwarf birch tundras, in which two microgroups are clearly distinguished, are common on the second terrace above the flood plain. The microrelief of these tundras is characterized by rather large hillocks, 0.2-0.4 m in height and 0.5-1 (1.5) m across, associated with the growth of the soil cover. A Cetraria—Aulacomnium—lingenberry—dwarf birch microgroup occurs on hillocks. The seasonal thaw beneath this microgroup reaches a depth of 0.3-0.5 m. The soils are swampy tundra permafrost soils, peaty and gleyey. In the seepage zones between hillocks, on swampy tundra permafrost gleyey soils, is found a sphagnum—cotton grass microgroup, beneath which thawing extends to a depth of 0.7-1 m.

Flat—hummocky cloudberry—Ledum sphagnum—cladonia peat bogs are very characteristic also of the second terrace of the Novyi Port region. Lichen—cotton grass—Ledum microphytocenoses of the tussocks and sphagnum—cloudberry microphytocenoses of the flat areas between tussocks and channels along frost cracks, differing in depth of seasonal thawing, have been described for the tussocky surfaces of hummocks.

The mosaic structure of the plant cover is clearly expressed also in variherbaceous—shrub tundra in the Noril'sk region. Restriction of this association to certain soil conditions was discussed previously by the author (Moskalenko, 1965). On the slopes of hills on the second terrace of the Noril'sk River, one may frequently encounter oxytropis—willow—Dryadaceae tundra. On such tundra two microgroups may be clearly distinguished, associated with different elements of the microrelief, the origin of which is due to frost cracking and erosion. The microrelief of the oxytropis—willow—Dryadaceae tundra is characterized by hillocks 0.2-0.4 m high and 2-3 m across, separated by channels and swales. The hillocks in this tundra are the sites of a lichen—astragalus oxytropis—Dryadaceae microgroup, and the depth of seasonal thaw-

TABLE 1. Indicator Scheme for the Novyi Port Region

Physiognomic component of the landscape	Indicated decipient component	
	Lithology of the soil-forming material	Depth of seasonal thaw, m
I. Treeless tundra, covered in shore zone by willow thickets and sedge meadows, with ramparts occupied by shrubby and mossy tundra, with moss—Ledum and sedge—sphagnum communities in flat swampy areas		
1. Herbaceous willow thickets, sedge and cold-tolerant sedge meadows along the shore	Sands	0.5-0.8
2. Complex of moss—Ledum communities in polygonal ground and sedge—sphagnum communities in seepage areas	Sandy loams and sands with layers of loam, covered by peat 0.25-0.3 m thick	0.25-0.7
3. Sedge—shrub tundra (ridges)	Sands	0.9-1.2
4. Cotton grass—shrub mossy tundra (ridges)	Sands	0.5-0.9
5. Sedge—sphagnum communities (swales)	Peat 0.3-0.4 m thick, underlain by sands or sandy loams	0.4-0.6
II. Flood plain of the Pya—Syadéi—Yakha with ramparts covered with shrubby tundra and meander swales occupied by mossy and sedge communities, with willow thickets near the channel		
1. Grass— and sedge—shrub tundra (ridges)	Sands	0.9-1.4
2. Sedge— and cotton grass—shrub—moss and moss—sedge communities (swales in flat areas)	Sands	0.4-0.9
III. 1st terrace with moss—shrub tundra on drained areas, with flat—hummocky peat bogs on flat swampy areas, with "khasyreis" covered with sedge—sphagnum communities		
1. Moss—shrub tundra	Sands and sandy loams	0.7-1
2. Complex of lichen—Ledum communities of hillocks and cloudberry—sedge—sphagnum communities of channels and seepage areas (flat—hummocky peat bogs)	Peat	0.3-0.6
3. Sedge—sphagnum communities, bordered by moss—willow—dwarf birch ("khasyreis")	Sands with layers of loam	0.5-1.1
IV. 2nd terrace with patchy shrub, lichen, and moss tundra at the terrace edge, a central hillocky area with lichen—moss shrubby and moss—herb—shrub tundra, with flat—hummocky and polygonal thermokarst peat bogs on swampy tracts and "khasyreis" covered with sedge—sphagnum and shrub—moss communities		
1. Patchy grass— and sedge—shrub and shrub—lichen tundra (at edge of terrace)	Sands with layers of sandy loan and loam	0.9-1.6
2. Moss—cotton grass—shrub and sedge— and grass—shrub mossy tundra (along edge of terrace)	Sands and sandy loams	0.3-0.9
3. Lichen—moss cotton grass— and sedge—shrub tundra (central surface of terrace)	Sands and sandy loams with layers of loam	0.7-1
4. Moss—herb—shrub tundra (central part of terrace)	Loams and sandy loams, underlain by sands	0.5-0.9
5. Sedge and sedge—sphagnum communities, on higher polygonal tracts shrub—moss (flat surface of "khasyreis")	Loams	0.3-0.7
6. Complex of lichen—Ledum communities on hillocks and moss—sedge—sphagnum communities in channels and seepage areas (flat—hummocky peat bogs)	Peat	0.3-0.7
7. Complex of lichen— and moss—Ledum communities in polygons and sedge—sphagnum communities in seepage areas (polygonal—thermokarst peat bogs)	Peat 0.25-0.5 m thick, underlain by sands	0.3-0.7

ing beneath them is 0.9-1.15 m. Moss—dwarf birch—willow microgroups are found in the swales and channels, and the depth of thawing beneath these zones is 0.5-0.7 m. Patches free of soil cover are frequently found, in which the deepest seasonal thawing is observed, amounting to 1.2-1.4 m.

The microgroups of oxytropsis—willow—Dryadaceae tundra are clearly identified on air photos taken in the spring when the tundra preserves its brown prevegetation aspect. At this time the Dryadaceae tundra gives a reticulated wavy banded figure on the photo. Microgroups of hillocks give a gray tone, and the microgroups in the channels and swales between hillocks are medium to dark gray. In summer, when the leaves of summer species are out and the tundra has become green, the differences in tone of the photo images of microgroups diminish, and it is appreciably more difficult to distinguish them. On photos taken in the summer, one may clearly distinguish the light tone of medallion-like patches commonly found in the oxytropsis—willow—Dryadaceae tundra, and because of which the air photos of such tundra usually have a speckled or mottled appearance.

In the Noril'sk region mosaic structure is also clearly expressed in peat bogs found on the hillocks of swollen segments of the second terrace. These peat bogs have a distinct polygonal pattern, clearly seen on air photos. In the polygons one may distinguish microgroups of tussocks and of spaces between hummocks, indicating different depths of seasonal thaw. The peat bogs of the Noril'sk region differ considerably in structure from the flat—hummocky peat bogs of the Tyumen' Oblast, examined above, but in species of plants they are rather similar (coefficient of floristic community according to Zhakkar, prepared for peat bogs of the different regions, is 42%), and the microphytocenoses distinguished on the Noril'sk peat bogs indicate the same depths as the microgroups on peat bogs of the Novyi Port region.

We have thus examined a series of mosaic structures in different parts of northern Siberia, due to inhomogeneities of environmental conditions in the different elements of microrelief and nanorelief, appearing either as a consequence of cryogenic phenomena (hillocks, swales) or as a result of the activity of plants (tussocks). Since microgroups restricted to definite relief elements have been studied, it is advisable to use as indicators not the taxonomic units of the plant cover (microassociations) but the physiognomic features of landscape units, called microfacies.* The more so that, from our observations, microgroups on a particular element of microrelief preserve their indicator significance in different plant communities within a single tract. Thus, the Ledum—lingenberry—bearberry—cladonia microgroups of hummocks indicate cryptopodzol soils on the surface of the first terrace in the Pur region, also in the thin larch—Siberian pine Ledum—bearberry—lingenberry forests, and also in thin larch—Siberian pine dwarf birch—lingenberry woods or copses.

The monotypic character of microgroups in different plant communities of forest-tundra was noted by Norin and Rakhmanina (1963). By using microclimatic observations, these workers showed that temperature differences of the surface in the upper layers, and also the moisture contents of different nanorelief elements in a single community, are greater than the differences in microclimate of identical nanorelief elements with different types of plant cover. This allows microgroups in structurally invariant species to exist in very different communities.

In all probability physiognomic components of microfacies may indicate not only soil varieties or depth of seasonal thaw, as shown above, but also moisture content of the soil and, possibly, some of the physicomechanical and chemical properties of the soil.

---

* By microfacies we mean the smallest elements of horizontal subdivision of the landscape not capable of further subdivision without changes in vertical structure, but possessing definite completeness.

In future studies of microphytocenoses for indicator purposes, it seems advisable to focus attention on solution of the following problems: a) discovery of features permitting us to fix the boundaries of the mosaic structure associated with inhomogeneity of the environment and due to peculiarities of the plants themselves; b) evaluation of the different methods of studying microgroups for distinguishing the best for indicator purposes; and c) explanation of possible interpretation of microphytocenoses on air photos and of extrapolation of their indicator significance. In the north, the characteristic structure of the image of the landscape indicator on large-scale air photos is related to differentiation of the plant cover by microrelief elements. Field interpretation of microfacies here permits one to explain the air-photo image of indicator facies.

LITERATURE CITED

Andreev, A. V., and Stal', Z. S., "Geologic structure and geomorphology of the left side of the Pur River along its lower course," Trudy VNIGRI, No. 158 (1960).

Ivanova, E. N., "Some structural patterns of the soil cover in tundra and forest-tundra along the coast of the Gulf of Ob," in: Soils of the Urals and of Western and Central Siberia, Izd. AN SSSR, Moscow (1962).

Lavrenko, E. M., "Microcomplexity and mosaic structure of the plant cover of steppes as a result of the activity of animals and plants," Trudy Bot. Inst. im. V. L. Komarova AN SSSR, Seriya 3 (Geobotanika), No. 8 (1962).

Levina, F. Ya., "Complexity and mosaic structure of vegetation and the classification of complexes," Bot. Zhurn., Vol. 43, No. 12 (1958).

Mel'nikov, E. S., Abramov, S. P., Goryainov, N. N., Laptev, V. F., Myaskovskii, O. M., and Tagunova, L. N., An Engineering-Geological Survey in the Region of Permafrost. (Experiment in Working Out Integrated Methods for Northwestern Siberia), Data of the 7th All-Union Interdepartmental Conference on Geocryology (Permafrost) (Materialy VIII Vses. Mezhduved. Soveshch. po Geokriologii (Merzlotovedeniyu)), Moscow (1966).

Minyaev, N. A., The Structure of Plant Associations, Izd. AN SSSR, Moscow–Leningrad (1963).

Moskalenko, N. G., "The plant cover in the vicinity of Noril'sk," Bot. Zhurn., Vol. 50, No. 6 (1965).

Norin, B. N., "Complexity and mosaic structure of the plant cover of forest-tundra," in: Problems of Botany, Vol. 6, Izd. AN SSSR, Moscow–Leningrad (1962).

Norin, B. N., "Community complexity of the plant cover of forest-tundra," Bot. Zhurn., Vol. 50, No. 6 (1965).

Norin, B. N., and Rakhmanina, A. T., "Interrelations of microclimate and the structure of plant cover in forest-tundra," Bot. Zhurn., Vol. 48 (1963).

Orlova, V. V., "Western Siberia," in: The Climate of the SSSR, No. 4, Gidrometeoizdat, Leningrad (1962).

Petrovskii, V. V., "Complex associations in the plant cover of tundra and forest tundra," in: Problems of Botany, Vol. 6, Izd. AN SSSR, Moscow–Leningrad (1962).

Sakharov, M. I., "Elements of forest biogeocenoses," Dokl. Akad. Nauk SSSR, Vol. 71, No. 3 (1950).

Viktorov, S. V., Vostokova, E. A., and Vyshivkin, D. D., Introduction to Indicator Geobotany, Izd. MGU (1962).

Yaroshenko, P. D., "Microstructure and macrostructure of the plant cover," Trudy Azerb. Fil. AN SSSR, Vol. 5, Baku (1934).

Zhukov, V. F., and Saltykov, N. N., Exploration of the Northern Port in the Gulf of Ob, No. 3, Izd. GUSMP, Moscow–Leningrad (1953).

THE COINCIDENCE OF GEOBOTANICAL
AND GEOMORPHIC INDICATORS
ON THE STEPPES OF TRANSBAIKALIA
T. N. Moiseeva

The purpose of the present work is examination of the coincidence of different ecological variants of certain steppe communities (as seen on the steppes of Transbaikalia) and relief elements. The article also seeks explanation of the possible use of these systematic combinations of relief and vegetation for indicator purposes.

Indicator investigations ever more and more clearly evolve from purely geobotanical and landscape investigations. It is usually shown (Viktorov, 1966) that in landscape investigations both vegetation and relief are used jointly as indicators, since they are the most readily observed physiognomic components of the landscape. However, though a close connection between vegetation and relief is widely recognized, there has thus far been little study to discover whether the connection is invariable, so that both components may be used for indicating as a single complex, jointly one with the other.

Nichols (1917) suggested that we call a relief element with its covering vegetation a "physiographic unit." Viktorov used this concept for indicator purposes (1947, 1966). Thus, the coincidence of boundaries of relief elements and certain units of the plant cover is assumed to be somewhat *a priori*. Meanwhile, in works on landscape science we encounter numerous statements concerning the possible inhomogeneity of the plant cover within a single, even very small, element of relief. Therefore, analysis of the character of the connection between vegetation and relief and of the possibility of their joint use as indicators must be considered an urgent problem in the theory of landscape indicators.

Investigations undertaken by us for the particular study of this problem were carried out on the steppes of the Kyra region of southern Transbaikalia, in the Tarbal'dzhei, Kurultyken, Barun-Khargaruk, Ust'-Ilya, and other areas. The steppes here are partly lithophilous, rich in petrophytes, partly true feather-grass steppes, and partly meadow-grass steppes.

For simplification of the problem, those community-indicators have been chosen as objects of study that are most frequently employed on steppes for determining moisture and rubbly character of the soil: petrophytic and mesophytic communities.

Analysis of the literature convinced us that *a priori* support of the combination of vegetation and relief elements into a single landscape entity will not be very convincing so long as the investigator works with rather large units of the plant cover, such as formations and even classes and groups of associations. This was shown very graphically by Dokhman (1954) for the vegetation of Mugodzhar, i.e., a region with extensive steppes having abundant rubbly and moist meadow-grass soils, and similar to the region of the present work in its landscape relations. Only very detailed analysis, with study of the ecological features of each association, based on consideration of the abundance of individual species, permits us fully to uncover the relation of vegetation to relief and the underlying rock or soil. For this we must turn to a study of the ecological features of different associations. These features may be most clearly delineated by fluctuations in abundance of particular species, primarily species belonging to different ecological groups. After we thus describe the ecology of the different associations, we may then advance to a study of the relations between these associations and the geomorphic conditions.

The first step of our work was thus discovery of the ecological trait of the association by computing the abundance of representatives of different ecological groups in the various associations. In doing this we turned our attention to three ecological groups most closely related to the contemplated indicator objects. Such groups are xerophytes, petrophytes, and mesophytes. Representatives of the first group are the principal plants distributed over the steppes and calculations of them were not made. Petrophytes, according to existing ecological representatives, should be found in association with rubbly tracts, mesophytes with moderately moist areas.

In separating species into ecological groups, we used as our base the data on ecology found in the book "Flora of the SSSR." These ecological data for individual species were refined by various sources from the literature and, in part, by handbooks dealing with other parts of the SSSR. As a result of this treatment of widely distributed species, forms included as petrophytes were Patrinia rupestris (Pall.), Polygonum angustifolium Pall., Leucopoa albida (Turcz.) v. Krecz., Thymus sibiricus L., Saxifraga spinulosa Adams; those listed as mesophytes were Trifolium lupinaster L., Sanguisorba officinalis L., Geranium Wlassovianum Fisch., Atermisia sericea Web.; and those considered to be xerophytes were Stipa decipiens P. Smirn, Festuca ovina L., Tanacetum sibiricum, Artemisia frigida Willd., and Poa botryoides Trin. After this step, we treated field descriptions (in all we had 68 descriptions available). The treatment was begun with a determination of the degree of petrophytic or mesophytic character of a given test area. This determination was made on the basis of number of species of a given ecological group having the greatest abundance on the scale of Drudé. For example, if two species were present in an association with an abundance of sol. and one species with an abundance of sp., the degree of petrophytic character was expressed by 1 sp. As a result of this treatment, each description obtained two conventional signs, expressing the greatest abundance of petrophytes and mesophytes and the number of species with this maximum appraisal. These evaluations appeared as abbreviations of the following type: P_1 sp., M_2 sol., P_3 sol, M_1 cop., and so forth.

In further discussions all descriptions, in keeping with the designations obtained, have been arranged according to degree of petrophytic or mesophytic character. This arrangement offered some difficulties in the case when a species was evaluated sol.-sp., sp.-cop., and so forth.

In the final consideration, we distinguished three degrees of petrophytic and mesophytic character. The first petrophytic degree (P_1) embraced descriptions in which petrophytes were represented by the evaluations sol.-5 sol. and the evaluations 1 sol.-sp.-2 sol.-sp. The second degree (P_2) included descriptions with evaluations 3 sol.-sp.-5 sol.-sp, 1 sp.-6 sp. The third degree (P_3) included descriptions with the evaluations 1 sp.-cop.^{3}-5 sp.-cop.^{3} and 1 cop.^{1}-5 cop.^{3}, i.e., all subdivisions of degrees of "cop." Degrees of mesophytic character exhibit the same range (M_1, M_2, and M_3).

After the associations are arranged according to degrees of petrophytic and mesophytic character in keeping with their floristic composition, by the geomorphic data in the descriptions, the position of the given sample area is analyzed for conditions of relief (exposure and steepness of slope) and for distribution of areas on the upper parts, middle parts, or lower parts of slopes, at the foot of a slope, or in a depression between slopes.

Below we deal with some of the results of comparing positions of particular associations on the scales of petrophytic and mesophytic quality with geomorphic conditions.

One of the most widespread types of steppe in the region of our investigation is the litho-philous steppe, represented by formations of the tansy (Tanacetum sibiricum) steppe. On the whole the ecological range of this latter type is very broad, embracing all petrophytic and mesophytic steppes and being found, thus, in very different habitats.

Relief features within this formation have also proved to be very inhomogeneous. The slope ranges from 1 to 25°. Areas of the given formation have been noted on all parts of the slopes (even at the top and in the middle segments) and are not present only on the floor of the basins. No distinct dominance of certain exposures was observed.

Analysis relative to individual associations gave more definite results. In one of the most frequently encountered associations – fescue–tansy (Festuca ovina–Tanacetum sibiricum) – correlation was detected only with average and high degrees of petrophytic character and chiefly only with a low degree of mesophytic character. This association was noted exclusively on the upper and middle parts of the slopes, about as commonly on the one as the other. The slopes are rather steep. Of nine sections on which the fescue–tansy community was described, two had slopes from 10 to 15°, three from 15 to 20°, and three of 20° (one section was found on a high flat saddle). Slopes with eastern exposure predominate; southern exposure is somewhat less common. The association was not observed on slopes with northern exposure.

The lespedeza–tansy association (Lespedeza hedysaroides (Pall.) Kitagawa) is somewhat similar to that described above. It also occurs where the degree of petrophytic character is P_2 and P_3 and the degree of mesophytic quality M_1 and M_2. The dominant habitat of this association is the upper parts of slopes (four cases out of seven) where slopes are appreciable. Of the described seven sections only one was on a slope of 10 to 15°; three sections lay on slopes of 15-20°, and three on slopes steeper than 20°.

A somewhat less definite ecological niche is occupied by the koelaria–tansy association (Koelaria gracilis Pers.–Tanacetum sibiricum), which was found on slopes of 15-20° as well as on slopes of 1-5°. It still resembles that community described above, however, since the steeper slopes predominate and there is a clear correlation between the association and the middle part of the slope.

The feather grass–tansy association (Stipa decipiens–Tanacetum sibiricum, with Stipa sibirica sometimes replacing S. decipiens) exhibits a constant correlation with the lower slopes and the foot of slopes. Slopes where this association is found are no more than 10°, decreasing locally to 1-2°. The association belongs to communities of the first degree of petrophytic char-acter, but is found in equal numbers in degree M_1 and M_2. No constant relationship between the feather grass–tansy association and exposure on slopes was detected.

Thus, in petrophytic character and steepness of slope on which it occurs, this community is completely opposite to that described earlier, being rather weakly petrophytic and reaching the highest mesophytic degree. This follows from its relationship to favorable conditions of moisture, at the foot of gentle slopes.

In coming to some results of examination of the tansy steppe, we may state that of the number of communities in this formation, such associations as the fescue–tansy and lespedeza–tansy form stable and readily recognized combinations with relief elements, creating definite landscape facies. These combinations may be used for indicating decipient components of the landscape, soil in particular. Thus, the fescue–tansy and lespedeza–tansy communities on steep slopes, having a large content of petrophytic components, may be considered an indicator of rubbly soil, whereas the feather grass–tansy association indicates a soil with finer constituents. The koelaria–tansy association does not form such systematic combinations, but may enter as a component in very different landscape facies. It is therefore not usable as a landscape indicator (although individual species in it may have definite significance, but for purely botanical indication, not for landscape indication).

In this manner we analyzed the relationship between relief and several other communities (for which we had at least three descriptions). We thus succeeded in discovering that variherbaceous communities with dominance of burnet (Sanguisorba officinalis), situated exclusively in the third mesophytic degree, may occupy a wide range on the petrophytic scale, from P_1 to P_3, and is found on the flat bottoms of basins as well as on the upper parts of rather steep rubbly slopes, if these slopes have favorable moisture conditions. Thus, here there is no necessary correlation between vegetation and relief, and the burnet communities with variherbaceous components may be found in very different facies. The feather-grass community with Stipa sibirica (L.) Lam. and abundant herbs (Pedicularis flava Pall., Gypsophila dahurica Turcz., Hemarocallis minor Mill., Polemonium coeruleum L., and Iris ruthenica Ker. Gawe) also prove to be somewhat irregular on the ecological scale (P_1-P_2, M_2-M_3) and in regard to relief. On the whole they favor the lower slopes, commonly found in the middle of these. The angles of slope range from 10 to 17°. The cold wormwood (Artemisia frigida) community, occurring entirely within P_1M_1 is an example of a very distinct coincidence with relief, since it is found either on the floors of basins or on the lower slopes, on slopes not exceeding 10°.

Thus, the tendency to form a stable combination with relief is found chiefly in those communities that have the least range in petrophytic and mesophytic character. This leads us to assume that landscape indication (meaning by this integral indication by coincident elements of plant cover and relief), despite its incontrovertible effectiveness and advantage over other forms of indication, is not always possible, because of the existence of some communities with rather broad ranges of geomorphic conditions of their habitat. We therefore think that progress in landscape indication does not exclude geobotanical indication based on ecological analysis of floristic lists and on interpretation of variations in abundance, viability, and thickness of species possessing rather definite indicator significance.

LITERATURE CITED

Dokhman, G. I., The Vegetation of Mugodzhar, Geografgiz, Moscow (1964).

Nichols, G., "The interpretation of certain terms and concepts in the ecological classification of plant communities," Plant World, 1917.

Viktorov, S. V., "Geobotanical regionalization as a method of geologic investigation," Byull. MOIP, Otd. Biol., Vol. 52, No. 2 (1947).

Viktorov, S. V., The Use of Geographic-Indicator Investigations in Engineering Geology, Nedra, Moscow (1966).

THE USE OF INDICATORS FOR PREDICTION
IN STUDYING THE EVOLUTION
OF SOME DESERT LANDSCAPES
M. T. Ilyushina

One of the urgent problems in indicator research is improvement of predictive elements. Both geobotanical indicators and landscape indicators (developed from the first) have displayed great success in determining by means of indicators various hard-to-observe components of the landscape in their equilibrium state.

At the present time increasing attention is being turned to the indication of processes. The most important of these and, at the same time, the most difficult is predictive indication, i.e., the interpretation of indicator data for constructing hypotheses concerning the appearance of certain processes or the future development of processes already begun.

In the present article we have tried to examine the predictive indication as applied to prediction of landscape evolution of ancient alluvial–deltaic plains and, in passing, to touch on some general questions on the use of predictive indication. Since the system of ideas on predictive indication has been but weakly developed at the present time, it is clearly advisable to introduce and explain some terms and definitions to be used below.

Among the cases of predictive indication it is advisable to distinguish: 1) true predictive indication and 2) retrospective predictive indication. In the first we have to do with prediction in its purest form, i.e., by using a definite composite natural circumstance as an indicator, we predict some process that has not yet begun. In the second case we speak of a process that has already begun, is already taking place. In analyzing the past phases of this process and the situation these phases lead to, we predict the future course. Prediction here is combined with retrospective consideration of past phases of the process, whence we obtain the name for this type of predictive indication. The views discussed here are a further development of views discussed by the Moscow Group of Specialists on Landscape-Indicator Research (Viktorov, 1967), whose purpose was refinement of the concepts associated with predictive indication.

True predictive indication and retrospective predictive indication differ rather strongly in the nature of investigation. In true predictive indication analysis of the interrelations between different landscape components is of greatest significance (Isachenko, 1953), and in some cases analysis of correlation between entire landscapes is utilized. In retrospective landscape indication comparison of different past stages of the process is the leading consideration, along with consideration of the physiognomic changes that may be used as indicators and formulation of predictions on the basis of such retrospective analysis.

The described types of predictive indication are different also in degree of reliability of the predictions obtained. In true predictive indication, even the trend of the future process is not altogether clear, and the reliability of prediction may be confirmed only by reference to similar predictions made for other regions that are similar to the investigated region in some measure. Retrospective prediction is much more reliable because the development of past stages of the process permits us to determine the trend and in some measure to forecast its future course.

The very means of indication in true and retrospective prediction are not completely alike. For true prediction we start from a definite landscape; and the characteristic features of its structure, commonly having a purely static character, we interpret as possible indicators. We introduce the element of dynamic interpretation ourselves, but we do not build it upon the initial data. In retrospective predictive indication, we have to do with an ecologic-genetic series, incomplete but already in progress to some extent (Dokhman, 1936), which we use as an indicator. In agreement with views developed earlier in indicator research (Viktorov, 1967; Vostokova, 1967), the best results are achieved when we use not only ecological-genetic series of plants as indicators, but also the more widely understood landscape—genetic series. In analogy with ecological-genetic series, we mean the change of landscape units in space that reflect their genetic preeminence in time.

Despite the certain differences, true and retrospective predictions are closely related. It is necessary to predict changes of processes, even those that have not begun (i.e., to predict in the direct meaning of the word), without having determined how processes that have begun will develop further (i.e., without consideration of the results of retrospective prediction). Predictive indication in the widest sense of the word is therefore impossible without combining it with another of the discussed types of prediction. On the whole, the scheme of predictive indication may be subdivided into the following stages: 1) study of past stages in development of processes taking place in the investigated region and analysis of the landscape-genetic series created by them for evaluation of the tendentious course, character, and rate of the processes; 2) interpretation of the landscape-genetic series created by the processes in order to predict their future course (retrospective predictive indication); 3) examination of different variants of interrelations that may arise between the processes taking place and a new superimposed process, the course of which is the result of prediction; and 4) selection of the most probable variant of the effect of the newly predicted process on the background of existing, already developed processes (true predictive indication).

In concluding these general considerations of predictive indication, we must emphasize its composite landscape character. At the present time we distinguish geobotanical, geomorphic, and landscape indication (Viktorov, 1967). With the first we take the plant cover as the indicator; with the second we use relief, and with the third we consider the whole complex of physiognomic elements of the landscape. Complete prediction is possible only in the latter case, since for this it is necessary, in the highest measure, to consider all combinations of conditions existing in the investigated region, inasmuch as it is impossible to conceive of such a process having some one landscape component that would not in some measure change the others (Polynov, 1952).

As the object for treating questions of predictive indication, we have selected part of the ancient alluvial—deltaic plain of Dzhana Dar'ya. This choice was dictated by the purely applied objectives of the investigation in which the author participated and also by consideration of the fact that the alluvial—deltaic plains (Dzhana Dar'ya, Kunya Dar'ya, and others) are presently the sites of extensive reclamation efforts, and the prediction of different processes taking place in these regions is therefore important for the national economy. In particular, as an example of predictive indication, we consider an attempt to predict the evolution of the landscape of the Dzhana-Dar'ya plain under the influence of water passing along abandoned channels of the Syr Dar'ya, of selective flooding and irrigation by means of artesian water. The object of our indication thus becomes the future aspect of the Dzhana Dar'ya plain after its flooding and irrigation.

The Dzhana Dar'ya plain lies chiefly within the Kzyl-Orda Oblast of the Kazakh SSR, in the region of abandoned Syr Dar'ya channels: Dzhana-Dar'ya, Kuvan-Dar'ya, and Inkar-Dar'ya. We investigated in detail part of the Dzhana-Dar'ya plain from the ruins of Kum-Kal Fortress to the ruins of Chilik-Rabat, a total area of about 6000 km², and made reconnaissance surveys (chiefly by observations from a plane and by landing the plane at certain selected localities) of adjacent regions.

The investigated region is an ancient alluvial—deltaic plain cut by ancient channels of the Syr Dar'ya: Dzhana-Dar'ya, Kuvan-Dar'ya, Inkar-Dar'ya, and others. In landscape-physiognomic relations, it represents a great massif of takyrs (clay deserts) and takyr-like surfaces along which are scattered individual isolated masses of sands of various sizes and thickness. Against the background of the alluvial plain occur remnants of Neogene deposits: the Uigarak, Sengir-Tam, Ak-Kyr, and Togyzken Plateaus. On the south the plain is bordered by the Kyzyl-Kum sands. Hydrogeologists have discovered artesian water from a deep aquifer underlying the entire Dzhana-Dar'ya plain. The water is fresh, with a mineralization on the order of 1 g/liter. The number of wells drilled into the producing aquifer is very large. Proposals have been made for using the flowing well-water both for creating reservoirs to guarantee water for abandoned pasture land and to provide irrigation near the wells.

The soil—plant cover of the plain exhibits a dominance of _Anabasis salsa_ (C.A.M.) Benth on the takyr-like sierozem (gray soil) loamy plain and of communities of wormwood and psammophytes on the sierozem sands of massifs composed of ancient alluvial—sand layers. Below we shall return to examine in more detail the vegetation in connection with the question of indication.

Indicator investigations on the Dzhana-Dar'ya plain were conducted earlier on areas east of the present detailed studies. These were water-indicator and rock-indicator studies of the All-Union Aerogeological Trust. Results were published (Vostokova, 1959; Vostokova and Savel'-eva, 1962). As a result of these investigations, Vostokova explained the presence on the plain of a series of ground covers and local accumulations of groundwater in lenses. The distributions of both depend most directly on relief of the impermeable cap of the aquifer, which is a gypseous Paleogene rock underlying a thick sequence of alluvial—deltaic sediments. The most important indicators of water are the different types of black crowfoot.

As a result of rock-indicator investigations (Vostokova and Savel'eva, 1962), three groups of plant communities were discovered, one of which (chiefly black wormwood and tamarisk) is associated with ancient alluvial sands, another (white wormwood and psammophyte) with Neogene sands, and the third (halophyte and shrub) with saline pseudosands.

Some indicator data are also found in the works of Pel't (1951), who traced the stages of plant change on ancient irrigated soils.

As seen from the above summary, past indicator investigations on the Dzhana Dar'ya plain had been devoted chiefly to geobotanical-indicator studies of statistical character. They did not touch on indication of natural processes in general or, much less, on predictive indication. Some data for prediction might be obtained only from the work of Pel't, but they touched only on areas modified by past activities of man. Thus, predictive indication for this region has yet to be worked out.

The first stage of our work was retrospective predictive indicator examination of the principal processes now developing on the Dzhana-Dar'ya plain. In agreement with the earlier discussion regarding retrospective predictive indication, we used the modern analytical method for developmental trends of landscape-genetic series. In order to determine the trend of development of this series, it was necessary first to discover the series. Discovery of the series, in turn, involved two stages: a) determination of the landscape series, i.e., the systematically recurring spatial changes of landscapes, and b) referral of some of these series to landscape-genetic series (i.e., to those in which the change of landscapes in space reflects the relations in time) and others to series genetically independent.

We discovered the landscape series by ordinary methods of profiling. For this we prepared schematic profiles many kilometers long, reflecting the general regional patterns of distribution of different landscapes in addition to smaller, but very detailed, profiles characterizing the structural details of boundary belts between different landscapes.

Generalized profiles of the plain (not touching the Neogene remnants) led to the discovery of two landscape series. One of these was related to massifs of sand (the study was made through the sand masses of Munarakuduk, Kel'te, Shakyr-Kazgan, Mullaly, and others). This series consisted chiefly of the following links: a) landscape of barchans and large hills of sand in combination with large deflation basins and a dominance of pioneer plants, psammophytes (species of the genera Calligonum, Ammodendron, Aristida, Corispermum); b) landscapes of indurated gently hummocky sands with a dominance of southern wormwood communities (Artemisia herba alba Asso) and saltwort (Salsola arbuscula Pall. and S. rigida Pall.) communities with numerous small deflation centers with psammophytic pioneer plants; and c) landscapes of fixed sands similar to those described above but with no small deflation centers.

It was noted that, in the distribution of all elements of the series mentioned, the fixed sands with small deflation areas were always in the transitional zone between winnowed and fixed sand. Landscapes of winnowed sands form isolated island sections amidst sand that is better fixed. In the deepest deflation basins groups of phreatophytes appear: large woody specimens of Haloxylon aphyllum (Minkw.) Iljin, Alhagi pseudalhagi (M.B.) Desv., Acroptilon picris (Pall.) Fisch. and Mey. If such basins are disposed within sands (such as the Kel'te sands), the thickness of phreatophyte plants (especially black crowfoot) is considerably less than where the phreatophytes occur in basins about the margin of sands (such as in the Shakyr-Kazgan sands). Sometimes one may observe zones, about the margin of the sands, in which black crowfoot patches are dying, having apparently been rather thick in the past (judging from the dimensions and thickness of dead material). Such zones were encountered by us along the margin of the Mullaly sands.

Another series was observed on the takyr plain, outside the sand massifs. It is multiple and consists of numerous alternations of takyr segments, completely free of plant cover and any recent detritus, takyrs overgrown in some measure with anabasis salt marshes (Anabasis ramosissima Minkw.) or other species of anabasis (Anabasis elatior, A. eriopoda (S.A.M.) Benth.), and, lastly, takyrs in some measure covered by shifting sand, sometimes forming thin miniature massifs with dominance of wormwood and saltwort.

After discovery of these generalized series it was necessary to decide whether the members of the series were genetically independent of each other or if we had to do with landscape-genetic series. We resolved these questions a) by means of studying the boundary zones between different members of the series, b) by searching for relics of one member of the series in neighboring areas, c) by analysis of information concerning the distribution of the different elements of the investigated series in the past. With the first method we started from the assumption that the transition from one element of a series to the next will more commonly take place with a genetic relation between the elements than without. Study of the material from this point of view showed that not only the individual elements of the series grade one into another very gradually, through smooth transitional zones without sharp boundaries, but both series — the series in sands and that in takyrs — grade rather imperceptibly one into the other. This transition is such that the margin of a sandy massif gradually merges with thin mantles of sand on takyrs near the sands, and these mantling sands thin gradually, little by little giving way to bare takyr surfaces.

However, a single gradual transition cannot be considered incontrovertible proof of a genetic relationship. We have therefore attached great significance to the discovery of relics of
one element of the series in an area occupied by another element. We noted many times small
relict patches of unwinnowed sands on the slopes of deflation basins and small fragments of un-
drifted sands of takyrs in mantles of sand covering takyr surfaces. This leads us to conclude
that the sands were previously more tightly bound, and that winnowing by the wind has taken
place recently, not yet having succeeded in removing all traces of past induration.

The formation of sandy mantles on takyrs also clearly represents a rather young process
superimposed on the takyrs, which have been open and clear until recently. One's attention is
drawn to the massive death of anabasis by drifting sands and to the appearance of individual
wormwood and saltwort, taking root not in the takyr but in the drift sand.

In order to evaluate the relations among the different elements of the described series
in the past, we used data from the literature and analyzed maps and air photos for different
years. It was found that in 1859 the Dzhana-Dar'ya plain was studied by the engineer A. N.
Gren. In his published account (Gren, 1863) no mention was made of drifting sands on the
takyrs, and descriptions of sands and takyrs are completely separate. Pel't (1951), who visited this region, also said nothing concerning any appreciable amount of drifting sand on the
takyrs. Lastly, on topographic maps drawn from 1956 air photos, areas of sand and amount
of drifting sand on the takyrs proved to be in many places much less than on air photos of 1966.

All the data discussed above, both landscape-morphological and historical−geographic,
attest to the existence on the Dzhana-Dar'ya plain of a process of gradual encroachment of
sands onto the takyr plain. The rate of this process (judging from a comparison of the data
of Gren and Pel't with recent information) has been rather lively, and if no external causes
retard it, we may predict reliably that in the next few decades the plain will be converted to
a patchy mosaic of small takyrs and sandy areas with a dominance of the latter. The mechanism involved is the accumulation of sand, passing across a takyr, at anabasis shrubs, and the
formation of a primary rather thin mantle of sand (seeded then by wormwood and saltwort),
growing, merging, and forming small masses of sand.

Another process requiring retrospective predictive indication is the formation of lacustrine salt-flat landscapes near wells from which flowing artesian water issues. Indication of
the course of this process is not complex, since it began recently and has not yet acquired any
complex stage of dissection. The collection of data reduces to landscape-ecological profiling
of the areas near the wells. The profiles are arranged along radii emanating from the well.
As a result of processing a number of profiles it was found that a rather constant landscape-
genetic series exists about each well. The initial link is a pond that forms near the well. In
the immediate vicinity of the well, a monotypic community of reed grass (Phragmites communis
Trin.) forms in the shore zone of the pond. On shores somewhat farther from the well, where
water is scarcer, a narrow belt of salt marsh (no more than 5 m wide) is developed, beyond
which there usually appear extensive meadows of saltmarsh grass (species of Aeluropus) with
shrubby saltmarsh tamarisk (Tamarix hispida Willd.) and tangled patches of annual saltwort
and sea blite.

Salt-flat meadows covered with saltmarsh grass merge with the surrounding takyrs, for
which, in addition to their typical anabasis formations, scattered groups of halophilic phreatophytes are present − Kalidium foliatum (Pall.) Moq., Halostachys caspica (Pall.) C.A.M., and
Nitraria Schoberi L. − pointing to increasing salinity of the groundwater beneath the takyrs.
This belt of salinized takyrs around wells has a width up to several hundred meters. Beyond
it lie typical takyrs, covered with anabasis, with no indications of solonchak degradation.

Interpretation of this series is rather simple, since all changes of communities in space uniquely indicate gradual salinization of takyrs because of uncontrolled flooding. However, at those wells where lucerne (alfalfa), strongly transpiring woody plants, and other phreatophytes have been planted, salinization of the takyrs has not been observed, since excess moisture is required for the transpiration. This is the situation at the Manekei, Myntai, and other wells.

These were the results of retrospective predictive indication at wells. They permit us to predict without doubt the process of progressive salinization of takyrs through flooding of artesian waters. With prolonged unregulated flow, swollen and damp salt flats form about the wells.

In going now to the next phase of predictive indication, to an examination of the different combinations of present-day processes described above and future flooding of the plain and passage of water through ancient channels and abandoned ditches of old irrigation systems, we may note several possible variants of these combinations. Introduction of large masses of water into the landscape of the plain must clearly retard movement of sand onto the takyrs. The cause of this is the high moisture contents of all habitats, including the sands, which will resist wind action, on the one hand, because of the moist sand, and will become favorable sites for growth of plants to bind the sand, on the other. Communities of phreatophytes about the margins of the sands become actively stimulated, and this will prevent spreading of sand across the plain. This is true, however, only if the water passes through channels near the sands. Masses of sand farthest from channels will remain unconsolidated and may remain sources of drifting sand and filling of water courses.

The interrelations between the future passage of water across the plain and the processes of solonchak formation at wells is more complex. In the early years, when all the irrigation systems of the plain are not fully operating, there may be a rather extensive dumping of irrigation water into the lowermost parts of the plain. The waters of flowing wells spread toward these same low areas. The two processes are additive in the early stages of utilization, and this may lead to intermittent, but appreciable, growth of solonchak. To avoid this, it is necessary to carefully prepare the region in advance for utilization, to create areas ready to receive and use the introduced masses of water. With expansion of areas for agriculture, the unproductive expenditure of water is curtailed and the intermittent growth of solonchak will cease. An essential type of preventive measure against salinization must be organization of proper irrigation at the wells, stoppage of uncontrolled flow, construction of reservoirs to be filled with water as it is needed. An important factor also is preservation of phreatophyte communities along channels through which water is passed. The strong evaporation of phreatophytes, especially black crowfoot, somewhat lowers the water table and prevents salinization along channels. From this point of view, the extensive growth of black crowfoot for fuel along the Dzhana-Dar'ya must be recognized as deleterious. In areas that have been completely worked out, where the crowfoot has already been destroyed, various halophytes commonly appear: saltmarsh grass, solonchak tamarisk, Halostachys. Continued reduction of black crowfoot undoubtedly increases salinization. Protection is also offered by growths of ditch reed at lakes that form near wells and along the Dzhana-Dar'ya banks, playing a role of active evaporation of moisture.

The picture of future landscape evolution on the Dzhana-Dar'ya plain, according to predictive indication, may be summarized in the following: 1) the greatest danger during irrigation and flooding of the plain is the possibility of salinization; 2) salinization will develop by superposition on solonchak degradation of low areas, already begun near flowing wells, of salt accumulation by dumping of overflow water from the ditches; 3) the process of sand migration onto the plain is continued only in zones receiving little water; 4) the best protection against these processes is better preservation of all natural phreatophytes in the landscape and even increasing them by planting woody varieties and sowing phreatophytic agricultural crops (this will preserve the plain from rise of the water table, utilize overflow water, and eliminate possible development of salinization zones.

All these measures must be carried out in the first stages of utilization at the same time as the first passage of water through the ditches, and the preparation for them must be begun even earlier. After full utilization of the plain, the necessity for many of these preventive measures may decline.

LITERATURE CITED

Dokhman, G. I., "Experiment in ecological-genetic classification of vegetation on the Isha forest-steppe," Byull. MOIP, Otd. Biol., Vol. 45, No. 3 (1936).

Gren, A. N., Expedition for Investigation of the Dzhana-Dar'ya River (1863).

Isachenko, A. G., Basic Questions of Physical Geography, Izd. Leningr. Gos. Univ. (1953).

Pel't, N. N., Ancient Irrigated Land on the Dzhana-Dar'ya Plain, Izv. BGO, No. 3 (1951).

Polynov, B. B., "Geochemical landscapes," in: Geographical Works, Moscow (1952).

Viktorov, S. V., "Indication of natural processes," in: Program, Summaries, and Annotated Reports on Problems of Plant Indicators, Leningrad (1967).

Vostokova, E. A., "The effect of buried relief on the plant cover on the Dzhana-Dar'ya plain," Izv. Akad. Nauk SSSR, Seriya Geogr., No. 3 (1959).

Vostokova, E. A., "Hydrogenetic ecological series of vegetation of desert regions," Zemlevedenie, Novaya Seriya, Vol. 8, Izd. Moskovsk. Gos. Univ., Moscow (1967).

Vostokova, E. A., and Savel'eva, V. A. "Vegetation as an indicator of genetically different sand deposits in the northern Kyzyl-Kum," in: Geobotanical Methods of Hydrogeological and Engineering-Geological Investigations, VSEGINGEO, Moscow (1962).

EXTRAPOLATION OF INDICATOR SCHEMES
WITHIN SALT MARSHES
M. A. Monakhov

One of the branches of landscape-indicator investigation that has been but incompletely worked out is the problem of extrapolation of indicator patterns, i.e., the problem of extending indicator schemes developed for some specific region to another region that is similar in some measure.

The theoretical aspects of extrapolation have been considered more than once (Vostokova, 1961; Vinogradov, 1962, 1964). Extensive data on the method of extrapolation has been discussed in a special collection – Air-Photo Standardization and Extrapolation (1967). in which the point of view of workers at the Laboratory of Aerial Methods is presented. However, direct study of the limits of applicability of different indicator schemes has been made on very few objects. Furthermore, until the factual aspect of possible extrapolation has been studied, many theoretical questions remain unresolved.

In the present article we consider the possibility of extrapolation of landscape indicator schemes in some types of salt flats. The choice of this distinctive object is dictated by considerations of two kinds. On the one side, the problems of indication of salinized tracts are of great practical significance in reclamation projects. On the other, because of the narrow ecological range of salt-flat vegetation and the extreme living conditions with which it is associated, it may be expected that the correlation of vegetation with habitats in quite different landscapes will be rather uniform. This appreciably simplifies the extrapolation and permits us to begin studying it with something less than the most complex variants, a procedure that seems best from the viewpoint of method.

Of all the varieties of salt-marsh regions, we have restricted ourselves to two: a) salt-marsh coasts of seas and lakes, and b) salt marshes (or salt flats) of interior-drainage basins. For both types of landscapes there is an extensive domestic literature, permitting us to construct rather rigid indicator schemes. In this construction, we have used for characteristics of salt-flat surfaces, with some modifications, the classification of Viktorov (1966), who distinguished armored (with a thick layer of salt) hummocky wet salt marshes and solid incrusted salt flats.

In generalizing the numerous studies of the shores of the Caspian and the Aral Sea (Il'in, 1927; Geideman and Doroshko, 1931; Kotov, 1928, 1931, 1938; Abolin and Chechkina, 1934; Benasevich et al., 1934; Pobedimova, 1935; Beideman, 1951, 1957; Egorov, 1954; Tagunova, 1957, 1960, 1961; and Katyshevtseva, 1960), it is possible to distinguish the following basic types of salt marshes along shores and to point out the indicator significance of each for a definite type and character of salinization.

I. Wet, incrusted or armored by the salt layer of the salt marsh or salt flat, barren of vegetation.

II. Wet salt marsh with glasswort (Salicornia herbacea L.) and annual species of Suaeda (chiefly S. maritina (L.) Dumort). Salt is chloride, mostly sodium chloride (salinization is sometimes merely surficial, possibly diminishing appreciably with depth).

III. Hummocky salt marsh with Halocnemum (H. strobilaceum M.V.): tussocks of Halocnemum. The salt is chloride (but with considerable sulfate content), chiefly sodium chloride (with appreciable content of calcium).

IV. Hilly or hummocky salt marshes, with hummocks about clumps of tamarisk (species of Tamarix), niter bush (Nitraria Schoberi L.), and boxthorn (Lycium). Chlorides and sulfides are almost equally represented in the salts, the probability of encountering one being about the same as the probability of encountering the other. This type of salt marsh is commonly combined with hummocky marshes with Halocnemum.

V. Salt-marsh meadows with a dominance of species of Puccinellia and Aeluropus. Salt is chiefly sulfate (in deep horizons possibly in equal parts with chloride or even subordinate to chloride).

Generalization of works on landscapes of interior-drainage basins (Artsimovich, 1911; Viktorov, 1955, 1959, 1960, 1966; Viktorov and Vostokova, 1956; Vostokova, 1961) permits us to distinguish a series of salt-marsh types completely similar to the series described above, and we may add two more types. One of these we shall call henceforth composite salt marsh (type VI), since it is a mosaic combination of patches of hummocky salt marshes, wet salt marshes, salt marshes with meadows and formations of plants that cease to be characteristic of salt flats but rather of solonets soils: Anabasis and Atriplex (chiefly A. canum S.A.M.). Salinization here is also of mosaic character, represented by complex alternation of sodium-chloride and calcium-sulfate zones. The other type of salt marsh (VII) we call the salt marsh-sand complex, since it consists of zones of salt flat and sands drifting over the flat. There is no clear correlation of this type of salt marsh with salinization, since the origin and character of the marsh may be highly variable. On the whole, however, this type is characterized by islands of less salinization against the general background of strongly saline substrate, and, in places, soil (sand) that is very poor in salt. The indicator scheme of salt flats of marine coasts and interior-drainage basins is fully made up by these seven, very heterogeneous types.

If we trace the limits of applicability of this scheme, they seem very broad (from northwestern Europe to South Africa). A series of works devoted to the salt marshes of Europe (Marsh, 1915; Wadham, 1920; Conway, 1933; Knight, 1934; Richards, 1934; Wiehe, 1935; and Hepburn, 1952) shows that the pioneers in colonizing marshes are glassworts, living in soils with high chloride salinization and readily persisting through flooding of marine waters. Communities of Suaeda maritima are also present here but are found somewhat farther from shore, on a soil undergoing some insignificant leaching. Halocnemum growths and chloride salt flats are also present here in combination with physiognomically similar communities of Suaeda fruticosa (L.) Moq., the indicator significance of which is not known. Salt meadows are very widespread, differing markedly in their specific plant forms from the salt meadows of the SSSR, but physiognomically and ecologically similar to them.

On the African coast of the Mediterranean we find many elements of our indicator scheme. In Libya, Wittchell (1928) and Willmott and Clarke (1960) have described maritime "sabhas" in the vicinity of Adjilat, and also between Misratah and Surt (Sirte): flat low areas filled with water in the spring but converted to wet or incrusted salt flats in the summer. According to Willmott the salt flats at the 'Ayn-Zar and Taguir "sabhas" are characterized chiefly by the dominance of sodium chloride.

In the central parts of depressions of the "sabha" type, vegetation is generally absent and one finds either solid salt flats with thin incrustation of salt or armored salt flats with a thick layer of salt. The plant cover about the margins of these lowlands, according to the above authors, is dominated by various species of glasswort (Salicornia herbacea L. and S. fruticosa L.).

One of the most completely investigated depressions of this type is the huge Tawurgha' "sabha." In articles devoted to this area, Berger-Landefeldt (1957, 1959) noted that glasswort and Halocnemum (in zones of chloride salinization) and salt meadows with mixed chloride-sulfate salinization are present, and that a huge field of incrusted salt is found in the center of the "sabha."

In this same depression occur extensive drifted sand on the salt flats, of weakly saline or nonsaline sands, creating numerous spits. These spits contain a distinctive ecological series from glycophilic communities of psammophytes on the crests of the spits to extreme halophytes at the base of the spits. This series strongly suggests the distribution of plant communities on Baer hills on the Tentyak-sor salt flat, until recently one of the bays of the Caspian Sea. In 1825 Éversman (1963) saw Tentyak-sor as a still shallow-water bay and first described the glycophilic psammophytic grasses (Agropyrum sibiricum (Willd) P.B. and others) growing on top of the hills. The ecological series later developed here has been studied in detail by Tagunova (1957, 1960). However, the physiognomic similarity of Tawurgha' and Tentyak-sor must clearly be considered merely as the convergence of two landscapes of different histories, since in the first we discover the process of recent burial of a salt flat by sand, in the second we observe the formation of young salt flats about hills that have long been in existence.

Hilly salt flats are also found on the coast of Libya (Desio, 1947). It is worth mentioning that phytogenic hills have formed here around artesian springs, the formation of which is associated with recent positive tectonic movements along the coast. The hilly salt flats thus prove to be concurrent (such as we pointed out for typical salt flats of the SSSR) with tectonic activity of the district and with regions of artesian water.

The shotts of Algeria and Tunisia are similar to the "sabhas" along the Libyan coast. In the Melrhir Shott, the one most extensively studied, we find both grasswort and Halocnemum communities and complexes of salt flats and sand. But in all sources known to the author, it was impossible to find data on salinization under these communities. Unfortunately this also applies to salt flats of the internal-drainage basins of Algeria, where Cannon (1913) noted the existence of broad areas of armored salt flats in the center of the depression and Halocnemum communities about the margin.

In the Somali desert, salt flats are associated chiefly with internal-drainage basins. In French Somaliland (Afars and Issas), de la Rue (1939) has found a rather large district of salt-flat desert in an interior-drainage basin near Lake Assal. In his sketch of the western shore of the lake, he showed a continuous mass of saline strata. Judging from the photographs illustrating de la Rue's work, the salt crust has a rough crumpled surface, but farther from the shore the crust shows large polygonal cracks.

The history of the Assal basin was described in detail by Joleaud (1927). He reported that the basin lay 150-160 m below sea level, was cut to a depth of about 10 m relative to the surrounding region, was 10-12 km across, and consisted of two subbasins. The eastern subbasin was filled with saline water, the western with a mass of salt. The shore of the basin was surrounded by a ring of gypsum accumulation from 50 to 500 m wide and about 5 m high. Two saline springs flowed into the lake, one of them warm, with a water temperature reaching 77°C. According to Joleaud, the basin had experienced a rather complex geologic history, having been subjected alternately to marine incursion (since it lay but 10 km from the sea) and to desiccation.

Halocnemum communities and growths of annual saltwort have developed in the Assal basin, covering chloride wet and incrusted salt marshes. There is no information concerning other landscapes at Assal. However, even with the poverty of information on this region, we may note at least three types of landscapes: armored chloride salt marsh, hummocky Halocnemum communities, and belts of sulfate salinization about the margin of the basin. Such features have been noted in a number of Ust-Urt depressions (Viktorov and Vostokova, 1956).

In the SSSR the greatest genetic similarity with the Assal basin is clearly found in the Ashchi-sor basin in Mangyshlak (Vyshivkin and Voronkova, 1962), which also experienced a number of marine incursions and was recently a marine gulf. The incompleteness of information on the landscapes of Assal makes it impossible to discover how complete this genetic similarity is in physiognomic relations.

Other basins in French Somaliland – Ganleh (also known as Galla) and Gobad – are covered almost entirely by a complex of salt marsh and sand. Information on the landscapes is somewhat contradictory. De la Rue (1939) mentioned extensive accumulations of sand, which he called dunes. In the marginal zone of the basin, especially on the northwest near Agueny, the sands have been smoothed off, are well cemented, and alternate with sulfate salt marshes. The origin of the sands is lacustrine. Vegetation is very sparse, consisting of small shrubs and semishrubs, both halophytes and psammophytes. Small groups of palms may sometimes be found. The same character of landscape is generally found in the Gobad basin, although no dunes are encountered. In Gobad, de la Rue noted lush development of tamarisk woods in the channels of intermittent streams in which underground flow was present.

The Italian investigator Braca (1939) described the Ganleh basin investigated by de la Rue (and called Galla by him) and considered it chiefly salt marsh formed in a zone of water accumulation from a number of mountain valleys surrounding the basin (such as the Ogag, Galafi-Daggira and other valleys). In Braca's descriptions there is no mention of sands in Ganleh. However, air photos supplied by him clearly show a combination of low sandy hillocks and salt-marsh segments in the landscape. The data of de la Rue given above therefore appear rather accurate.

These basins are thus examples of steady development of salt marsh and sand complexes (without participation of extensive areas of wet and incrusted salt marshes) with a mosaic alternation of different types of salinization. Similar depressions may be found in the SSSR, the most typical being the Assake-Audan (Viktorov, 1955), where bumpy sulfate salt marshes and drifted lacustrine sands form a complex mosaic of nonsalinized segments and areas in which sulfate salinization dominates.

In the Kalahari desert, salt marshes also form chiefly in internal-drainage basins. The largest of the Kalahari basins is the Makarikari. It is a huge gently sloping lowland, forming the reservoir for discharge of a number of streams, the largest of which is the Botletle. Streams falling into this lowland form extensive floods with a tremendous area of evaporation. Salinized clays are widespread in the Kalahari (Schultze, 1907). The introduction of salts, therefore, and their accumulation in the basin has been considerable, and this has led to the formation of great fields of armored salt flats.

A description of the salt-marsh basin of Makarikari given by Clifford (1929, 1930, 1938) permits us to recognize it as a zone of dominant development of wet salt marshes and oozes, the moisture of which is maintained by inflowing streams. These features are best developed in the parts of the basin called Soa (or Shua, also Nata) and Itvetve (also called Mokoamoto). On the other hand, near the mouth of the Botletle, solid surfaces of salt are found, with rare associations of shrubs and some salinized soils (similar to salt-marsh meadows).

According to Rey (1932), in the marginal parts of the basin are found bumpy chloride hummocky salt marshes, very similar physiognomically to the hummocky Halocnemum deserts of the SSSR, but floristically different.

At Namib, saltwort and Halocnemum meadows occupy extensive areas in dried-up coastal lagoons. The centers of the lagoons are generally occupied by viscous wet salt marsh. Salt-wort borders these centers, and salt-marsh lagoons occur on the higher zones between de-pressions (Schultze, 1907; Kaiser, 1923; Logan, 1960).

The above brief summary of salt deserts of marine coasts and interior-drainage basins is convincing in the great uniformity of the landscapes. Similarity appears not only in the gen-erality of types, but in the similar distribution in space. Thus, almost everywhere the lowest zones (the floor of an internal-drainage basin, the margin of a maritime marsh) is occupied by one variant or the other of dead salt marsh, barren of vegetation. The zone next to the lowest is occupied by Halocnemum or saltwort communities. In some places these communities, es-pecially in basins that are drying up, may be combined with remnants of shoreline vegetation: giant reed and bulrush. The next zone away contains either salt-marsh meadows or composite salt marshes. Both exhibit mosaic arrangement of salinization, but sulfates clearly tend to dominate. In our summary, this structure of salt marshes was noted in some degree of com-pleteness in Libya (the Tawurgha' "sabha"), in French Somaliland (Lake Assal), and in the schotts of Algeria. Some fragments of this scheme may be found also in very many other re-gions. During personal investigations in Iraq, we observed a very typical distribution of in-crustation salt marshes and hummocky Halocnemum associations about the marshes in the vicinity of Lake Habbaniya. A similar structure (with some substitutions of plants) has been discovered in some salt marshes of Australia at Lake Eyre (Madigan, 1930).

We should note that the different types of landscape indicators that we have distinguished do not have wholly identical limits of extrapolation. Wet, incrusted, and armored salt marshes, wet salt marshes with saltwort, hummocky salt marshes with Halocnemum, and hilly salt marshes have the greatest areas of extrapolation and they preserve their significance in all the regions examined. Salt-marsh meadows and composite salt marshes are much more vari-able in their indicator significance, and extrapolation of this part of the scheme is limited. We should probably seek the cause in the nearness of these last two types to interfluve areas with zonal vegetation, whereas there is almost complete independence of zonal conditions in the first three types.

The constant structure and appreciable floristic similarity of plant communities in the investigated salt marshes show that the indicator schemes when they contain rather widely rec-ognizable landscape indicators have the greatest possibilities of extrapolation. It is necessary to bear in mind, however, that in extrapolation we may come against the fact of replacement of some dominant type of indicator by another type, equivalent ecologically and physiognomic-ally, which does not essentially disturb the indicator scheme or alter its applicability.

Sometimes combinations of salt-marsh communities, even in regions far removed from each other, are practically indistinguishable. In the interior-drainage basins of southwestern Turkmenia, we observed belts of microcomplexes. The center of one of these microcomplexes was a wet chloride salt flat with no vegetation. Toward the margin of the basin we found a nar-row ring of saltwort and Halocnemum, and plots of salt-marsh grass (various species of Aeluropus). We observed very similar microcomplexes in the vicinity of Lake Habbaniya, in the region of the Ramadi upland, and at the Milh-Tartar salt marsh in Iraq. The number of such ecological analogies may be greatly multiplied (Chapman, 1960).

Thus, a broad potential opens up for extrapolation of indicator schemes for some types of salt flats in deserts and along coasts, improving the value of results obtained by indicator investigation.

LITERATURE CITED

Abolin, R. I., and Chechkina, A., "Vegetation of salt marshes, its use and adaptation," in:
      Problems of Plant-Growing Utilization of Deserts, No. 2, Izd. VIR, Leningrad (1934).

Air-Photo Standardization and Extrapolation (Method), Nauka, Leningrad (1967).

Artsimovich, V. S., Wet Salt Marshes in the Vicinity of Baskunchak Lake, Transactions of the
      Society of Naturalists at Kharkhov University (Trudy Obshch. Isbyt. Prirody pri Char'-
      kovskom Univ.) (1911).

Banasevich, N. N., Zonn, S. V., Kazmina, T. I., and Makkaveev, N. I., Salinization and Desalin-
      ization of Soils in Connection with Groundwater, Its Salinization, and the Effect of the
      Caspian Sea, Transactions of the Leningrad Branch of the All-Union Institute of Fertiliz-
      er and Agronomy (Trudy Leningr. Otd. Vses. Inst. Udobrenii i Agropochvovedeniya), No.
      29, Leningrad–Makhachkala (1934).

Beideman, I. N., "Change in plant cover along the shores and floor of Kirov Gulf in connection
      with withdrawal of the Caspian Sea," Bot. Zhurn., SSSR, No. 1 (1951).

Beideman, I. N., "Observations on vegetation changes along the coasts and salinization of sea
      water during recession of the Caspian Sea," Trudy Bot. Inst. AN SSSR, Seriya III, Geo-
      botanica, No. 11 (1957).

Berger-Landefeldt, U., "Beitrage zur Ökologie der Pflanzen Nordafriks, Salzpfannen Vegetatio,
      Vol. 7, fasc. 3 (1957); Vol. 9, fasc. 1-2 (1959).

Braca, G. G., Rilevementi Topografici; La Somalia Française e Africa Italiano, 3, L. Universo
      (1939).

Cannon, W., Botanical Features of the Algerian Sahara, Washington (1913).

Chapman, V. I., Salt Marshes and Salt Deserts of the World, New York (1960).

Clifford, B. A., "A journey by motor lorry from Mahalapye through the Kalahari Desert,"
      Geogr. J., Vol. 73, No. 4 (1929).

Clifford, B. A., "A reconnaissance of the great Makarikari Lake," Geogr. J., Vol. 77, No. 1
      (1930).

Clifford, B. A., "Across the great Makarikari Salt Lake," Geogr. J., Vol. 91, No. 3 (1938).

Conway, W. M., "Further observations of the salt marsh of Holm-next-the-sea, Norfolk,"
      J. Ecol., No. 2 (1933).

De la Rue, Aubert, La Somalie Française, Paris (1939).

Desio, A., "Geologia et morfologia," in: Il Saharo Italiano, Roma (1947).

Egorov, V. V., Formation of Maritime Salt Marshes on Marshy Terraces in the Western
      Caspian Region, Transactions of the V. V. Dokuchaev Soil Institute (Trudy Pochv. Inst.
      im. V. V. Dokuchaeva), Vol. 44 (1954).

Éversman, É. A., "Nature journal, written during expeditions to view the region between the
      Caspian and Aral Seas in 1825," in: First Russian Scientific Investigations of the Ust-Urt,
      Izd. AN SSSR, Moscow (1963).

Geideman, T. S., and Doroshko, I. N., "Sketch of the vegetation on the Sal'yani steppe," Trudy
      po Geobot. Obsled. Pastbishch SSR Azerbaidzhana, Seriya A, No. 8, Baku (1931).

Hepburn, J., Flowers of the Coast, London (1952).

Il'in, M. M., Vegetation of the El'ton Basin, Transactions of the Main Botanical Garden (Trudy
      Gl. Bot. Sada), Vol. 26, Leningrad (1927).

Japp, R. N., and Johns, D., "The salt marshes of the Dovey Estuary," J. Ecol., No. 5 (1917).

Joleaud, L., "La genèse de gisements de potasse d'apres les conditions géologique du lac Assal,"
      Rev. Scient., No. 9 (1927).

Kaiser, E., "Abtragung und Auflagerung in der Namib, der Sudafrikan Küstwüste," Geol. Charak-
      terbilder, No. 27/28 (1923).

Katyshevtseva, V. G., Change in Plant Cover on the Northern Coast of the Caspian Sea, Trans-
      actions of the Institute of Botany of the Academy of Sciences, Kazakh SSR (Trudy Inst.
      Botaniki AN Kazakh. SSR), Vol. 8 (1960).

Knight, B., "A salt-marsh study," Amer. J. Sci., Vol. 28 (1934).

Kotov, M. I., "New data on the vegetation of Arabatskaya bar," Zhurn. Russk. Bot. Obshch., 3-4 (1928).

Kotov, M. I., "Vegetation near the Sivash of the Over'yanovskoe salt lake and its shores," in: Anniversary Collection [Dedicated to] Academician B. A. Keller, Voronezh (1931).

Kotov, M. I., "Origin of the solonets and salt marsh complex in Transcaucasia," Sov. Botanika, No. 3 (1938).

Logan, R. F., "The Central Namib Desert," Nat. Acad. Sci. Nat. Res. Council, Washington (1960).

Madigan, C. T., "The Lake Eyre, South Australia," Geogr. J., Vol. 76, No. 3 (1930).

Marsh, A. A., "The maritime ecology of Holm-next-the-sea," J. Ecol., 3 (1915).

Pobedimova, E. G., Vegetation of Coastal Deserts and the Sands of Kara Bogaz, Transactions of the V. V. Dokuchaev Soil Institute (Trudy Pochv. Inst. im. V. V. Dokuchaeva), Vol. 2 (1935).

Rey, C. F., "Ngamiland and the Kalahari," Geogr. J., Vol. 80, No. 4 (1932).

Richards, F. Y., "The salt marshes of the Dovey estuary," Ann. Bot., 48 (1934).

Schultze, F., Aus Namaland und Kalahari, Berlin (1907).

Tagunova, L. N., "The geobotanical regions of the northeastern coast of the Caspian Sea," Vestnik Moskovsk. Gos. Univ., No. 4 (1957).

Tagunova, L. N., "Relations of the soil-plant cover of the northeastern coast of the Caspian Sea to conditions of salinization and soil moisture," Byull. MOIP, Novaya Seriya, Otd. Biol., Vol. 65, No. 1 (1960).

Tagunova, L. N., Development of Plant Cover on the Northeastern Coast of the Caspian Sea (in Connection with Conditions of Salinization of Soil-Forming Materials), Author's abstract of candidate's dissertation, Moskovsk. Goz. Univ. (1961).

Viktorov, S. V., "What the study of the vegetation of Sarykamysh and Assake-Audan has shown," Priroda, Vol. 1, No. 1 (1955).

Viktorov, S. V., "Geochemical facies and the plant cover of the bitter saline oozes of Barsa-Kel'mes," Byull. MOIP, Novaya Seriya, Otd. Biol., Vol. 64, No. 2 (1959).

Viktorov, S. V., "Ecological series of plants in connection with conditions of salinization in the region of Tuz-Kan Lake (Kyzyl Kum)," Byull. MOIP, Novaya Seriya, Otd. Biol., Vol. 65, No. 2 (1960).

Viktorov, S. V., Use of Indicator Geographic Investigations in Engineering Geology, Nedra, Moscow (1966).

Viktorov, S. V., and Vostokova, E. A., "Plant cover as an index of salinization in interior-drainage basins of the Ust-Urt," Izv. Akad. Nauk SSSR, Seriya Geogr., No. 1 (1956).

Vinogradov, B. V., "Geographic patterns of distant extrapolation of interpretive features of landscape analogs," in: Application of Aerial Methods for the Study of Groundwater, Izd. AN SSSR, Moscow (1962).

Vinogradov, B. V., "Ecological compensation, substitution, and extrapolation of plant indicators," in: Plant Indicators of Soils, Rocks, and Groundwater, Trudy MOIP, Vol. 8, Moscow (1964).

Vostokova, E. A., "Phytogenic tussock indicators of artesian springs (concerning geographic substitution of indicators)," Byull. MOIP, Novaya Seriya, Otd. Geol., Vol. 66, No. 3 (1961).

Vyshivkin, D. D., and Voronkova, L. F., "Geochemical features of the Ashchi-sor basin and their reflection in the plant cover," Vestnik Moskovsk. Gos. Univ., Seriya V, Geogr., No. 2 (1962).

Wadham, S. M., "Changes in the salt-marsh and sand dunes of Holm-next-the-sea," J. Ecol., 8 (1920).

Wiehe, P. O., "A quantitative study of the influence of tide upon populations of Salicornia europaea," J. Ecol., No. 2 (1935).

Willmott, S. G., and Clarke, J., Field Studies in Libya, Rec. Paper, ser. 4 (1960).

Wittchell, L., Klima und Landschaft in Tripolitanien, Hamburg (1928).

INDICATOR SIGNIFICANCE OF PLANT MESOCOMPLEXES IN SANDY MASSIFS OF THE NORTHERN ARAL REGION

N. N. Darchenkova

The reclamation and intelligent use of broad regions of semideserts for the domestic economy in the northern deserts of Kazakhstan represent one of the most important objectives for development of animal husbandry in our country. Sand massifs are in this respect of special interest, since they furnish year-round pasture, particularly valuable in the winter and also commonly good for hay. Mostly these masses of sand are most promising in considerations of water supply. The problem of water supply is of primary importance in the rational utilization of these sands as pasture lands, and also in solving other problems of the economy (building of settlements and industrial installations). Solution of the problem requires scientific basis for studying the masses of sand and prospecting for sources of water.

The work of hydrogeological parties in masses of sand, however, is commonly complicated by the difficulty of transport through the sands, especially for drilling equipment. With the existing methods of hydrogeological mapping it is very easy to overlook small lenses of fresh water, which are commonly among the basic sources of water supply (Vostokova, 1961). Auxiliary methods of investigation therefore gain value, methods that permit one quickly and with little expense to make a reconnaissance evaluation of the region, and to furnish better direction for special investigations. Indicator geobotany and landscape-indicator methods satisfy these needs at the present time (Vostokova, 1961; Viktorov, Vostokova, and Vyshivkin, 1962; Vinogradov, 1964).

In carrying out investigations by the suggested methods on masses of sand in the northern Aral region, we concluded that it is advisable in mapping to take into account not so much the indicator significance of separate communities as the indicator significance of their systematic combinations in space, i.e., in their mesocomplexes. The present article is devoted to this question.

By a sand massif we mean a limited region of piled-up, drifted sand, bare or to some degree overgrown with plants (slightly or half overgrown), commonly including tracts of sandy plains and sand-free material (remnants of bed rock, zones of sandy-loam, loam, or clay plains, takyrs, salt flats, and the like). Examples of sand massifs are found in the northern Aral region: Bol'shie and Malye Barsuki, Air-Kyzyl, Taldykum, and others. The sand massifs are generally bordered by high "marginal" hills, 3-10 m high. These hills commonly terminate abruptly against the plains at angles of 20-30°.

The material making up the main area of the sand massifs is most commonly fine- and medium-grained sand, mostly well sorted and having good permeability, low water-raising capacity (capillary potential), and low specific retention (Gael' et al., 1950; Petrov, 1950; Yakubov, 1955). Because of the high permeability of the sand, there is no well-defined surface runoff; rainfall is completely absorbed. Having low specific retention (porosity), the sands cannot physically accumulate large quantities of water within them, and the moisture therefore settles downward under the influence of gravity. The low capillary potential guarantees preservation of the moisture (that has infiltrated downward) from upward capillary movement toward the surface and, hence, from evaporation. Because of these properties, and also because of the specific climatic and geological conditions, lenses of fresh water may form in the sand massives, either floating on salt water or resting on a near-surface impermeable layer (Kunin, 1963). Near-surface water has been noted almost everywhere in the sand massifs of the semideserts and northern deserts, but the depth varies, and the mineralization also varies. Districts where subsand lenses of water form are found where the sand massives are most weakly cemented or are bare (Kunin, 1959). The infiltration of meteoric water is a decisive factor for the sand massifs of the semidesert and many massifs of the northern deserts (Gael' et al., 1950; Vladimirov and Fedin, 1954; Ageeva and Bulgakova, 1955; Akmedsafin et al., 1963). In hummocky sands of the semidesert and northern deserts, lenses are found in deflation basins usually at shallow depth, perhaps 1-3 m. Because of the nearness of this water in the basins to the surface, appreciable loss may occur through transpiration and evaporation. In view of the mode of occurrence of the lenses in the surrounding masses of sand, the water usually leaks out. A connection between vegetation and these shallow lenses of water by virtue of deep root systems has been noted by many authors (Gael' et al., 1949, 1950; Vostokova, 1953, 1967; Beideman, 1962; Rachinskaya, 1964).

Investigations (Rachinskaya, 1966) have permitted us to determine the indicator significance of 23 of the most widespread communities in the sand massifs and the adjoining regions.

Below we have described briefly the indicator significance of those communities that most frequently form combinations with each other, creating mesocomplexes in semideserts.

1. Communities of oleaster (Elaeagnus angustifolia L.) and willow (Salix acutifolia Willd., S. rosmarinifolia L.) with (sometimes) dog rose (Rosa cinnamomea L.) in deflation basins, low areas between hummocky sands, along the margins of sand massifs, at the contacts of aquifers and impermeable layers. The sands are unconsolidated; water lies at a depth of (0.5)-1.5 (2) m and has a mineralization of 0.1-1.0 (2.7) g/liter. The ground has been leached of the easily dissolved salts.

2. Communities of Halimodendron (H. halodendron (Pall). Voss.) on slopes of basins, low areas between hillocks, on saddles between hillocks, and also at the margins of sand massifs and estuarine sands, on consolidated and unconsolidated sands. Groundwater is at a depth of (0.5)1-2(2.5) m, and mineralization is 0.1-3.0 g/liter.

3. Communities of bulrush (Holoschoenus vulgalis Link.) in those habitats occupied by the above communities, but most commonly in the sand massifs themselves, in basins, and only occasionally along the margins of the massifs, where these grade into sand or sandy loam plains. The sands are loose. Water is found at a depth of 1.0-2.0 m and is fresh (mineralization of 0.1-1.0 g/liter).

4. Communities of reed grass (Calamagrostis epigeios (L.) Roth.), sometimes with Dodartia (D. orientalis L.), in the same habitats where woody-shrub communities are developed, as described above. Water lies at a depth of 1.0-2.0 m and has a mineralization that ranges up to 1.0 g/liter.

5. Communities of feather grass (Lasiagrostis splendens (Trin.) Kunth.) with licorice (Glycyrrhiza glabra L.), wild rye (Elymus akmolinensis Drob.), reed grass, giant reed (Phragmites communis Trin.), and others along the margins of sand massifs, in basins on slightly undulatory plains, rarely in deflation basins in the massifs themselves, but also on slopes of upland areas (at contacts between sands and clays, where fresh water accumulates). The deposits

are variable: bound sands with layers of loam, sandy loam, locally with dense clays near the surface. Depth to water ranges from (0.5)1.0 to 2-3 m, and mineralization ranges from 0.1 to 2(3) g/liter.

6. Communities of giant reed with mesophytes on both the slopes and the floors of deflation basins and in low areas between hillocks, frequently found on slopes and at slope inflections on marginal hillocks of the massif where it grades into the surrounding region, and also sometimes forms growths along the shores of fresh-water lakes (in the latter case with tuberous bulrush, Bolboschoenus maritimus (L.) Pall., rush, Juncus gerardii Loisel, related forms, Heleocharis pauciflora L., and others), and about springs. The deposits are sands, sandy loams, sometimes loams. The water has a mineralization of 0.1-1.5 (up to 2-3.5) g/liter and occurs at a depth of (0.0)1-2(3) m.

7. Communities of wormwood (Artemisia arenaria D.C.) are widespread in hummocky sands on all elements of relief, on broad low areas amidst the sand, on slightly undulatory and gently sloping undulatory areas and sloping sandy plains. The deposits are loose sands (in going to the northern desert subzone, these communities become indicators of water, pointing to mineralization of the water up to 2.0 g/liter at a depth of 1.5-4.5 m (in low areas)).

8. Communities of giant rye (Elymus giganteus Vahl.) with other grass (Aristida pennata Trin.) occur in semiconsolidated and weakly bound sands on both crests of hillocks and other elements of the relief where loose sands are almost unbound by vegetation. Like the wormwood communities, this type becomes an indicator of water in the subzone of the northern deserts, pointing to water at a depth of 1.0-2.5 m in low areas, with mineralization up to 1.0 g/liter.

9. Communities of licorice are found not only in the sand massifs (floors of basins, low areas between hillocks) and along their margins, but also in low areas and basins on gently sloping undulatory plains. In the sands these communities point to water at a depth of 0.5-2.0 m, fresh (up to 2.0 g/liter), and on the plains they indicate a depth of water of 3.0-6.0 m.

10. Communities of spirea (Spiraea crenata C.A. May, S. hypericifolia L.), like communities of licorice, occur in sand – in basins, indicating fresh water with a mineralization up to 1.0 g/liter at a depth of 0.5-2.0 m, and in low areas·on gently undulatory plains and gently sloping uplands, indicating deeper water at 2.5-6(7) m, both fresh and brackish. This water is commonly perched.

All the communities described above, as well as oleaster and willow communities, form on soils leached of readily soluble salts.

11. Communities of feather grass with salt-marsh species. Some of these communities may contain white wormwood (Artemisia Lercheana Web. and Stechm.) and goosefoot (Kochia prostrata (L.) Schrad). They occur about the margins of salt-marsh basins. The deposits may be sands or sandy loams and loams. The water is mineralized (over 3.0 g/liter) and occurs at depths of 1.0-3.0 m. The ground is weakly salinized.

12. Communities of giant reed with salt-marsh species (dwarfed reeds) are found about the borders of salt-marsh basins, indicating sandy loams and loams with weak or moderate salinization, where saline water (3.0-17.0 g/liter) occurs at a depth of 0.5-1.5 (up to 2.0) m.

13. Communities of tamarisk (Tamarix ramosissima Ledeb., T. hispida Willd.) are found on hummocky sands, on sloping undulatory sandy plains, along the borders of salt marshes on sands and sandy loams. Water is brackish and saline at a depth of 1.0-2.5(4) m.

14. Communities of camel's thorn (Alhagi pseudalhagi (M.B.) Desv.) are found most frequently about the borders of salt-marsh basins and they grow on slightly undulatory sandy plains. The water shows spotty mineralization (from fresh to saline) and occurs at a depth of 1.0-3.0 m. The soil is sand and sandy loam.

15. Communities of wheatgrass (Agropyrum sibiricum (Willd.) Beanv.). Ephedra shrubs are sometimes found (Ephedra distachya L.). The communities are widely developed on hummocky sands on different elements of relief according to degree of consolidation and equally on sloping undulatory plains, where they cover considerable area. The soil consists of bound sand and sandy loam leached of the readily soluble salts.

16. Communities of psammophytic shrubs (the buckwheat Calligonum aphyllum (Pall.) Guerke and others) with Becker's fescue (Festuca Beckeri Hask.), koeleria (Koeleria glauca D.C.), found widely on sand massifs, most frequently in districts of weakly bound or semibound sands. The sands are always unconsolidated and have been leached of their readily soluble salts.

Communities òf black wormwood (Artemisia pauciflora Web.), Anabasis salsa (C.A.M.) Benth., Halocnemum strobilaceum (Pall.) M.B., and Salicornia herbacea L. are not found on the sand massifs, but without considering them it is generally impossible to give a full picture of all the processes taking place in the sands. We therefore note here that the first two communities indicate clays and loams with strong sulfate salinization. The next two are developed in salt marshes and indicate saline water at a depth of 0.5-1.5 m and soil that exhibits strong, commonly very strong chloride salinization, occasionally chloride–sulfate salinization. The soil consists of loam, muds, and muddy clayey sands.

In the sand massifs the hillocks and basins alternate constantly. The substratum is also observed to be constantly mobile, resulting in the formation of complex plant cover (Korovin, 1934, 1962; Gael' et al., 1949, 1950). In the practice of indicator investigations of the sandy massifs, therefore, presently used with medium- and small-scale photos, there is considerably more promise in mapping larger complex communities than areas of individual communities. However, the nature, structure, and type of these communities have remained until now almost unstudied, and before conducting indicator mapping it is necessary to make some brief characterizations.

In the sand massifs the complexity of plant cover, as pointed out above, is due to eolian processes, depth of groundwater, and other factors, which stand out as leading factors in various parts of the sand massif. We may see from the above data that each community-indicator is associated with specific elements of the relief. Thus, buckwheat communities and groups occur mostly on the crests of hills. Phreatophyte communities are most frequently found on the floors of deflation basins, where groundwater is nearest the surface. In the landscape of sand, different physiographic elements or physiographic units exhibit constant alternation. Under physiographic units the American investigator Nichols (1917) placed combinations formed by plant communities and the definite relief element on which they occur.

Systematic changes of relief in sand massifs and of other natural conditions associated with them lead to the formation of systematic combinations, called "mesocomplexes," meaning, by this, complexes due to the effect of mesorelief. "In mesocomplexes the elements of mesorelief play a role just as important as the elements of nanorelief and microrelief in patchy nanocomplexes and microcomplexes" (Levina, 1964);" ... the coexistence of different associations in the complex, associated with definite elements of mesorelief, ... is more logically called a mesocomplex" (idem). Below we use the term mesocomplex in just this sense, describing actual mesocomplexes of the semidesert zone in sand massifs of the northern Aral region.

Mesocomplexes formed by combinations of sand hills and deflation basins are developed in hilly or hilly-ridged sands by the effect of wind on the underlying layer of sand. Eolian processes lead to some rearrangement of the mechanical properties of the deposits. For example, where wind action has been greatest, sorting of the sand is greatest, the content of silty particles is minimal. The percentage of clay-sized particles is insignificant, as low as 3.5% (Chetyrkin, 1930; Gael' and Ostanin, 1939; Yakubov, 1941, 1955). Districts less subject to wind action have a somewhat higher percentage of clay-sized particles (up to 5-6%). According to the extent of wind action, a substratum on more consolidated sands begins to develop soil. The following mesocomplexes are observed according to extent of wind action on the surface and the mobility of the sandy substratum.

The crest of a sand hill frequently attacked by wind is occupied chiefly by groups or communities of psammophytes, among which pioneer plants, of premier value in binding the sand, are commonly dominant. These groups and communities are of giant rye, wormwood, buckwheat, aristid grass, Becker's fescue, and others. Projected cover does not exceed 20-30%. Sometimes crests that are less affected by wind are also more strongly bound by vegetation. In this case the dominant groups of associations are of wormwood (associations of wormwood–wheatgrass, wormwood, wormwood–grass, and wormwood–giant rye with buckwheat).

On hill slopes occur communities of the wormwood and fescue groups of associations. With a projected cover of 20-30% the dominant associations are fescue–koeleria, fescue–astragalus, and fescue, i.e., the group of fescue associations. The wormwood–wheatgrass, wormwood, wormwood–psammophyte associations are characteristic of consolidated and semi-consolidated slopes. The projected cover in this case increases to 60-70%.

Hill slopes either grade directly into the floors of basins or begin to grade into a basin slope. Transition to basin slope is commonly very poorly expressed in the relief and is marked or emphasized only by a change of association, so that the boundaries of relief elements are traced only by the vegetation. Most commonly, where hill slope grades into basin slope, the hill slope is well indurated and is characterized by development of the wormwood group of associations. The basin slopes are not well bound (except rarely) by communities of the fescue group of associations. The projected cover does not exceed 20-30%. The basin floor is also most commonly weakly bound by communities of psammophytes and has a projected cover of 20-25%. Fescue, fescue–astragalus, and fescue–psammophyte groups or associations are dominant.

The described mesocomplex due to growth of vegetation on a sandy substratum subjected to wind action also characterizes in part the mechanical constitution of sands in individual districts and the nature of the winnowing process. Communities and groups of buckwheat, fescue, giant rye, aristid grass, and others thus point to the best sorting of sand and the smallest percentage of clay particles (0.1-2%). Wormwood communities point to loose sands (alphitite up to 5%), but the wheatgrass group of associations is found chiefly on bound sand (alphitite up to 10%). Therefore, in analyzing the alternation of members of this complex, we may in some measure describe the eolian processes in sand massifs. Buckwheat, fescue, and giant rye groups and communities point to districts of crests, slopes, basin floors most subject to winnowing action of wind. The wormwood group of associations is developed on semibound and bound parts of the massif. The complex represented on hill crests of buckwheat associations (or groups) with giant rye, and aristid grass, with a projected cover of 10-20%, on slopes of wormwood–wheatgrass associations with a projected cover of 50-60%, and in the basin of fescue–astragalus community, characterizes districts of weakly bound and semibound sands, where winnowing by wind is dominant. The complex where we find weakly fixed basins and strongly fixed slopes and hill crests may point to the initial stages of winnowing, forming foci of deflation.

All basins in sand massifs are not characterized by the development of psammophyte communities. Many are occupied by phreatophyte communities because of the nearness of groundwater. In this case we observed a complex consisting of members of the preceding complex (elements of crests and slopes) and of phreatophyte communities. The following geographic variants may occur according to depth and abundance of water.

Subvariant I. On crests and upper slopes are developed communities and groups of psammophytes: buckwheat, wormwood, wheatgrass, fescue, woodruff, giant rye, and others. The projected cover is 20-30%. On hill slopes and basin slopes the wormwood group of associations is mostly present, with a projected cover of 10-20%. In many places the lower slopes have phreatophytes (giant reed, Dodartia, Halimodendron) in psammophyte communities, or a giant reed community develops. On basin floors phreatophyte communities dominate. The projected cover reaches 90-100%. Here we find communities of oleaster, willow, Halimodendron, feather grass, reed grass, and sandy bulrush. This member of the complex indicates fresh water in all cases, lying at a depth of 1.0-1.5 m. The complex is most typical of sand hills in semideserts, indicating a zone of partial discharge of a subsand lens.

Subvariant II. In the Taldykum sand massif, in the northern part of the Bol'shie Barsuki, one may observe not only phreatophytes on the basin floor, but also hygrophytes. The complex, from sand hills to the floors of low areas, consists of the following members. The flat tops of sand hills are covered by the fescue–psammophyte community. Crests grade directly into basin slopes, on which are developed phreatophyte communities: willow, dog rose, giant reed, reed grass, and others. On basin floors we find close herbaceous covers consisting of moisture-loving variherbaceous forms with abundant sedge (Carex Karelini Meinsh., C. gracilis Curt., Thalictrum minus L., Tanacetum vulgare L., Inula britannica L., Scutellaria galericulata L., and others). Along the edges of basin floors occur trees of oleaster and shrubby willow. The combination of these mesocomplexes points to areas of sand massifs where abundant groundwater discharges. This water is presently used as a source of water supply in the northern part of the Bol'shie Barsuki.

Mesocomplexes formed by combination of sand hills with segments of plain among the sands are developed in separate sections of hilly, slightly hilly, or low-hilly sands, in which alternating sand hills give way to broad low areas between hills, representing essentially sloping undulatory sandy plains in the sand massifs. We have observed such districts in the sand massifs of the Taup tract (Kokzhiek, Karakol', and others), in Sarbulakkumi, in the Air-Kyzyl massif, and elsewhere. The mesocomplex has the following character. On hilltops facing low areas between hills, we find buckwheat–psammophyte, buckwheat–wheatgrass–wormwood, wormwood–grass, and other communities from the wormwood and buckwheat groups of associations. The projected cover is usually 30-40%. On hill slopes the wormwood–grass community is dominant, commonly with ephedra. Sandy plains are dominated by the wormwood group of associations. The projected cover here is 60-70%. Districts of sand massifs with such mesocomplexes are most strongly indurated and least favorable for formation of fresh-water accumulations in the sand massifs. However, these mesocomplexes may be indicators of the leakage of groundwater that may accumulate in other parts of the massifs. In this case buckwheat–psammophyte and wormwood communities are developed on the crests and upper slopes facing plains districts. On lower slopes we find only wheatgrass–wormwood communities or these communities with phreatophytes (giant reed, Dodartia, and others). At the base of hills and on slope inflections, phreatophyte communities are dominant (giant reed, wild rye, licorice, and others), indicating fresh water at a depth of 1.0-1.5 m. This member of the mesocomplex gives way to the element described above, characteristic of plains districts, i.e., wormwood–wheatgrass communities. At some places where the slope bends, we may observe very abundant leakage of fresh water. Here we encounter growths of oleaster trees, Halimodendron bushes, tangled liana, and, among herbs, luxuriant meadow herbs of various types and meadow grass, i.e., these communities may be considered fragments of taiga vegetation.

Most commonly a sand massif will be well separated from the surrounding region by "marginal" sand hills of rather appreciable height. However, though the relief may exhibit a somewhat sharp boundary for the massif, the vegetation grades less perceptibly. Nevertheless, the mesocomplexes that form at the boundaries may display the effect of the sand mass on the adjoining region and may define the hydrogeological conditions for these outlying districts. The margins of the sand massifs studied by us generally represent a belt 100-200 m wide. On crests and slopes, especially upper slopes, communities appear that are characteristic of these elements within the massif itself, as described above. These are communities of the buckwheat and wormwood groups of associations. On the middle and lower slopes, wormwood associations predominate with considerable admixture of wheatgrass, and also with kochia, ephedra, and other related forms. At breaks in slope we find the beginning of slightly undulatory, sloping undulatory, and almost flat plains occupied by communities of the wormwood, wheatgrass, and the white wormwood group of associations. This member of the complex is developed only on sands with increased amounts of silty material and where the sands have become consolidated, or where sandy loams have formed. Some sand massifs grade directly into loamy plains, and a narrow belt of wheatgrass—white wormwood communities at the foot of the hill grades into a community of white wormwood, black wormwood, or anabasis. Such mesocomplexes indicate a rather rapid change from sandy material to heavier (denser) material (up to clay) and to less favorable conditions for accumulation of groundwater or its manifestation (seep, wedging out of lenses). Such places are not promising for winter quarters, since there is little probability of finding groundwater near the surface here.

The borders of sand massifs are sometimes very promising for dug wells, however. Where this is true we may note several subvariants of the mesocomplexes depending on depth and mineralization of groundwater.

Subvariant I. The mesocomplex is distributed in those districts at the margins of massifs where groundwater at the foot of a hill is at a depth of 2.5-3.5 m. On the crests and upper slopes of such hills, we find buckwheat—psammophyte communities. Down the slope occur wormwood—wheatgrass and wheatgrass—white wormwood communities. At the foot of the slope, on terrace surfaces, we find wheatgrass—feather grass—Dodartia community with kochia and white wormwood; i.e., ombrophyte communities develop, with phreatophytes. Here, in addition to feather grass, we may find giant reed and camel's thorn. The following member of the complex, appearing on the plain, is represented by the same communities of the wheatgrass and white wormwood groups of associations. It is sometimes represented by anabasis and black and white wormwood communities.

Subvariant II. This typical mesocomplex is observed most frequently along the margins of sand massifs. The members of the complex on crests and slopes are the same as in subvariant I. But on the lower slopes one frequently finds phreatophytes (giant reed, Dodartia, and others). At the foot of the hills at the break in slope to the plain, a belt some tens of meters across contains abundant phreatophyte communities of the oleaster and willow, Halimodendron, feather grass, giant reed, and other groups of associations. For example, we found the following associations: Halimodendron—feather grass—wild rye with willow trees, feather grass—giant reed—wild rye, feather grass—grass, oleaster—grass, willow—tansy, and others. In this complex, phreatophyte communities indicate groundwater at a depth of 1-2.5 m. The water is fresh with solid dry residue of 0.1-1.0 g/liter.

Subvariant III. The mesocomplex forms along the margins of sand massifs, where water occurs near the surface, leading to the formation of swampy areas or to the emergence of springs. In this complex, the tops of the hills, as in the preceding complexes, are covered by psammophyte communities, but along the slopes we find phreatophyte communities (giant reed, giant reed—willow, grass—variherbaceous, and others). At the foot of the hills occur swampy areas or very moist ground, where sedge—variherbaceous, willow—giant reed—mesophytic, variherbaceous, and other communities develop. Standing water at the surface is frequently observed. Sometimes fresh-water ponds are formed in such areas. The margins of these ponds are overgrown with bulrush, tuberous rush, juncus rush, heleocharis rush, and other similar plants. In a belt up to 500 m wide immediately next to this habitat, the dominant communities are grass—variherbaceous and meadow grass, giving way gradually to the complex growing on the plains. Such swampy zones with dominance of mesophytic variherbaceous plants have been noted by Gael' in the Bol'shie Barsuki; he has called them "karataly." We have observed such areas on the eastern margin of the northern part of the Bol'shie Barsuki, on the northern and eastern margins of the Air-Kyzyl massif, and on the northern and northwestern border of the Malye Barsuki.

Subvariant IV. In individual bordering zones the sand massif may grade directly into a salt marsh. At the transition from the foot of a sand hill, a terrace-like surface usually begins, sloping gently toward the salt-marsh basin. The basin is clearly separated from the terrace-like surface by a well-defined ledge or bench. The psammophyte communities described above dominate on the sands next to the salt marsh, on crests and slopes. On the lower slopes or at the foot of the sand hill we find Halimodendron—giant reed, grass-variherbaceous, giant reed, giant reed—camel's thorn, and other communities. The succeeding member of the complex is developed on the terrace-like surface between sand hills and salt marsh. Feather grass communities dominate here. Sometimes the communities of the base of the hill (Halimodendron—giant reed, giant reed, and others) are absent, and we find members of the complex developed on slopes grading directly into the feather grass belt. Toward the salt marsh, the feather grass belt gives way to meadow grass—salt-marsh communities or communities of dwarf ditch reed (giant reed), feather grass, occupying the lowest part of the terrace-like surface along the margin of the salt-marsh basin. At the brow of the slope of the salt-marsh basin, one may observe a very narrow band of variherbaceous salt-marsh plants. This belt may disappear, and meadow grass—salt marsh communities grade into the last member of the complex, glasswort and Halocnemum communities, which occur in the marginal zone of the salt marsh. This mesocomplex points to rapid increase in mineralization of the water leaking from the sand massif, and special attention should be given any considerations to use the zones for sinking wells for drinking water.

Mesocomplexes may be used not only as independent indicators, but also for defining hydrogeological conditions and rocks at individual points in the sand massifs and, even more, for characterizing large parts of the massif, sometimes for the entire massif. Thus, an almost continuous belt of mesocomplexes along the margins of the Air-Kyzyl sand massif points to a great ample zone of fresh-water seepage from subsand lenses in the massif, permitting us to describe the northern part of the Air-Kyzyl massif as promising for obtaining good drinking water.

The legend on a map distinguishing mesocomplexes will differ from the legend of a map prepared with consideration of indicator significance of individual communities. On the basis of the indicator significance of individual communities, we prepared a map of the Air-Kyzyl sand massif. Part of the legend for this map is shown in Table 1.

In the rest of our work we used types of mesocomplexes, and the legend on the map prepared with consideration of mesocomplexes had the following form: In column one we show the index designating our mesocomplex on the map (I, II, III, etc.). In the second column we show the name of the mesocomplex, reflecting both the relief element and the communities that combine in the mesocomplex. In the third column we give the subvariants as they were described above. The last column shows the indicator interpretation of the given mesocomplex, giving special attention to the characteristics of the processes indicated by the structure of the complex.

The practical significance of the mesocomplexes is thus found in the fact that the mesocomplexes represent a rather convenient unit for medium-scale indicator mapping. For example, actual indicators of water and, primarily, phreatophyte communities very commonly oc-

TABLE 1. Part of Legend for Mesocomplex Map

Index	Indicator features		Indicator conditions		
	Vegetation	Habitat, relief	Lithology of surface deposits	Depth of water table, m	Minerali-zation, g/liter
I	Wormwood–psammophyte communities with oleaster and willow communities	Hilly polygonal sands with deflation basins	Sands	0.5-1.5	up to 3.0
II	Wormwood–psammophyte groups with willow and giant reed	Ridged semiconsolidated sands	Sands	1-3	up to 3.0

cupy such insignificant areas that on medium-scale maps (which are most widely used in indicator studies) they are illustrated by dimensionless signs. This appreciably complicates the search for fresh or slightly brackish water and frequently reduces to the identification of points that separately are difficult to find. The construction of maps using mesocomplexes gives more direction to the search for water, since it orients the hydrogeologist toward the discovery of pattern in the plotted combinations of communities developed on successive elements of relief.

The use of mesocomplexes facilitates also indicator interpretation, since each mesocomplex is designated on air photos by a repetitive combination, defining patterns. Therefore, interpretation of water indicators, as an example, may be made not by small individual elements, commonly points, of a pattern created by groups of phreatophytes, but by different, easily distinguishable types of localities, which, in essence, are also mesocomplexes. There is no doubt about the convenience and effectiveness of mapping on the basis of analyzing the mesocomplex structure of the landscape.

LITERATURE CITED

Ageeva, N. T., and Bulgakova, N. B., Root Systems of Tree and Shrub Genera of the Bol'shie Barsuki Sands, Uchenye Zap. Kazakhsk. Gos. Univ., Biologiya i Pochvovedeniya, Vol. 17 (1955).

Akhmedsafin, U. M., Gubarev, A. N., Sadykov, Zh. S., and Yakupova, N. Ya., "Groundwater of the Turgai Plain and its reserves," Izv. AN Kazakhsk. SSR, Seriya Geol., No. 2 (1963).

Beideman, I. N., Bespalova, Z. G., and Rakhmanina, A. T., Ecological–Geobotanical and Land-Improvement Investigation in the Kura-Araks Lowland of Transcaucasia, Izd. AN SSSR, Moscow–Leningrad (1962).

Chetyrkin, A. S., The Air-Kyzyl Sands at Irgiz and Methods of Fixing Them, Trudy Turgaisk. Melior. Éksped. za 1920-1923-24 gg., Tashkent (1930).

Gael', A. G., Kolikov, M. S., Malyugin, E. A., and Ostanin, E. S., Sands of the Ural–Émba Region and Methods of Utilizing Them, Trudy Inst. Pustyn', Vol. 1 (1949).

Gael', A. G., Kolikov, M. S., Malyugin, E. A., and Ostanin, E. S., Sand Deserts of the Northern Aral Region and Methods of Utilizing Them, Trudy Inst. Pustyn', Vol. 2 (1950).

Gael', A. G., and Ostanin, E. S., "The southern Kazakhstan sand massif of Myunkum," in: Utilization of Deserts and Uplands, Moscow (1939).

Korovin, E. P., The Vegetation of Central Asia, Saogiz, Tashkent (1934).

Korovin, E. P., The Vegetation of Central Asia and Southern Kazakhstan, Izd. AN UzbSSR, Book 1 (1961); Book 2 (1962).

Kunin, V. N., Local Water of Deserts and Problems of Its Use, Izd. AN SSSR, Moscow (1959).

Kunin, V. N., "The significance of fresh-water lenses," in: Lenses of Fresh Water in Deserts, Izd. AN SSSR, Moscow (1963).

Levina, F. Ya., Vegetation of the Semideserts of the Northern Caspian Region and Its Value as Fodder, Nauka, Moscow–Leningrad (1964).

Nichols, G., "The interpretation of certain terms and concepts in the ecological classification of plant communities," Plant World (1917).

Petrov, M. P., Shifting Sands and Means of Controlling Them, Geografgiz, Moscow (1950).

Rachinskaya, N. N., "Vegetation of the borders of sand massifs and their indicator significance," Byull. MOIP, Novaya Seriya, Otd. Geol., Vol. 39, No. 5 (1964).

Rachinskaya, N. N., Geobotanical Indicator Investigations in Sand Massifs of the Northern Aral Region, Author's abstract of candidate's dissertation, Moscow (1966).

Viktorov, S. V., Vostokova, R. A., and Vyshivkin, D. D., Introduction to Indicator Geobotany, Izd. Moskovsk. Gos. Univ. (1962).

Vinogradov, B. V., Plant Indicators and Their Use in Studying Natural Resources, Vysshaya Shkola, Moscow (1964).

Vladimirov, N. M., and Fedin, N. F., "Conditions of forming mineralization of groundwater in the sand massifs of the northern Caspian region," Izv. AN Kazakhsk. SSR, Seriya Geol., No. 18 (1954).

Vostokova, E. A., Vegetation as an Index of Geological and Hydrogeological Conditions of Deserts and Semideserts in Connection with Their Utilization, Author's abstract of candidate's dissertation, Moscow (1953).

Vostokova, E. A., Geobotanical Methods of Prospecting for Groundwater in Arid Regions of the Soviet Union, Gosgeoltekhizdat, Moscow (1961).

Vostokova, E. A., "Hydrogenic ecological series of plants in desert regions," Zemlevedenie, Vol. 8, Izd. Moskovsk. Gos. Univ. (1967).

Yakubov, T. F., Sands of the Deserts of the Northern Caspian Region, Problems of Soviet Soil Science (Probl. Sov. Pochvovedeniya), No. 12, Moscow (1941).

Yakubov, T. F., Sand Deserts and Semideserts of the Northern Caspian Region, Izd. AN SSSR, Moscow (1955).

MAN-MADE INDICATORS AND THEIR SIGNIFICANCE IN LANDSCAPE-INDICATOR INVESTIGATIONS

S. V. Viktorov

With the development of landscape-indicator science, the range of indicators has steadily expanded. Primarily we have used plants chiefly as indicators, and this has furnished us with a basis for indicator geobotany. Later, views on indicators became more widely spread to other physiognomic components of the landscape: relief, surficial soil horizons. However, until now, it has been very rare that landscape elements due to the activity of man have been used as indicators. Nevertheless, these elements very commonly are clearly physiognomic, and, with sufficiently thorough study of their relations to various decipient conditions, they may acquire great indicator value.

By man-made indicators we mean those physiognomic elements of the landscape that have appeared as a result of man's activity and may be used as indicators because of their indisputable recognizability and their definite, rather stable relations to particular components of the physicogeographical environment.

The first and most fundamental question that arises during examination of this group of indicators is the question of the extent to which they are necessary and effective and the purposes for which they might be used. The rather restricted experience with man-made indicators that we have acquired in recent years convinces us that man-made indicators are most useful not in those regions that have been continuously used by man but in those regions where traces of his activity are scattered more or less sporadically and are associated with some definite conditions determining the possible record of man's activity at just those points. In regions where man has altered the natural conditions and replaced natural landscapes by artificial surfaces, the discrimination of man-made indicators is considerably more complex, although possible. Therefore, in this article we shall dwell only on the first variant of applying man-made indicators, since they are the most promising.

The use of man-made indicators in sparsely inhabited regions is based on the premise that people settle where they have found the most favorable natural conditions for their existence (good pasture for their herds, good soil for farming, building materials, water, fuel, etc.). Of course, these natural conditions have not always been decisive, and in the appearance of temporary stations and settlements significance also attaches to considerations of protection from attack, suitable commerce, and so forth, but, other conditions being equal, natural physicogeographical conditions have strongly influenced the disposition of settlement. Therefore, analysis of the distribution of districts where some traces of man's residence has been noted against a sparsely settled background region may have definite indicator value.

In order to be more concrete in our consideration of the possible practical use of man-made indicators, we have described their significance in the specific examples of deserts, semideserts, and steppes. In these types of regions, man-made indicators acquire practical

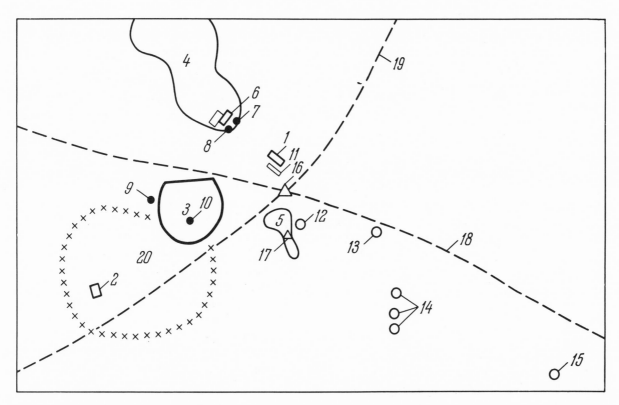

Fig. 1. Sketch of the ruins of Taleh. 1) Ruins of Taleh tower; 2) ruins of Falat tower; 3) ruins of Silsilat fortress; 4-5) garden plots; 6) quarries and ruins of Dar-Ilalo tower; 7-10) storehouses for grain; 11) water reservoir; 12-15) graves of Abdullah-Hassan, Soldan-Nura, and others; 16-17) springs; 18-19) caravan routes; 20) area with remains of building.

value as indicators of places that are promising for various improvement measures, irrigation, water for small-oasis farms; in other words, for complex animal husbandry and plant-growing. Among these indicators, primary place belongs to complex systems that may be called relict oases, i.e., traces of ancient oases once used by man but now deserted, marking, as it were, the former presence of fertile soil, drinking water, and other prerequisite features for utilization. Among such relics we may find some that are very old and some very young.

It is natural that the use of the oldest relics of oases for indicator purposes must be made with the greatest care, since they may be associated with ecological conditions so far back in the past that any indication of present conditions proves to be impossible. But oasis relics dating from a few centuries back may have definite interest for soil indication.

As an example of a young oasis relic we may point out Taleh in Somalia. Once there apparently existed here some small Somalian settlement, disappearing because of raiding tribes from the Danakil Desert and Ethiopia. During the struggle of the inhabitants of the former British Protectorate of Somaliland with the colonists, the leader of the rebels Mohammed-ben-Abdullah Hassan (known as the Mad Mullah) founded at Taleh a rather large settlement, making it his principal base. In February 1920 Taleh was almost entirely destroyed by the barbarous bombing of English planes (Jardin, 1923; Lewis, 1965; Battersby, 1914). Concerning the present aspect of Taleh, now completely abandoned by man, we may judge from the sketch in Fig. 1, prepared by Macfadyen in 1930 (Macfadyen, 1931) and somewhat modified by us. It clearly shows the remains of the Silsilat fortress (having complexly built walls with paths inside them), rooms, large storehouses, and also the large Taleh tower, considered completely impregnable, and the somewhat smaller Falat tower. Around these were scattered the ruins of quarries, watering troughs, viaducts equipped with stone and arched reservoirs, numerous graves and garden plots, probably used for domestic growing of gardens in the past. The largest of these is the Dar-Ilalo plot. Many of the ruins have been leveled almost to the ground and can be distinguished with difficulty even on air photos. This is especially true of the wall enclosing the cultivated areas. Despite the difficulty of recognition, the entire assemblage of Taleh remains, in the view of the investigator, a place suitable for utilization. According to travelers who have visited this place, there is plenty of drinking water and soil suitable for farming, although many springs have been buried beneath the crumbled arches of reservoirs.

Such oasis assemblages are also found in the deserts of the SSSR. For example, in 1966 the geobotanical party of the Central Asia Aerogeological Expedition, in which the author participated, described for the first time traces of buildings and an irrigation system in the gully network of Agach-Kuduk on the northwestern escarpments of Ust-Urt. Similar descriptions had been made even earlier by Vyalov (1934) in the Kainar gully on the western escarpment of Ust-Urt and by Nikitin (1893) in the Igindy-Bulak tract on Dzhil'tau Mesa north of Ust-Urt. In the northern Aral region, a few kilometers northwest of Perovskii Gulf, between it and the Kop-Bulak wells, one may see on the slopes of the plateau bordering the sea coast a large area with a well-developed system of irrigation ditches, traces of large buildings, water reservoirs, and enclosing walls (Ak-Kuduk district). Such man-made indicators are very numerous on the Kunya-Dar'ya plain near the ruins of the fortification surrounding ancient Khorezm: the At-Aka, Ak-Chigil' (Akich-Gelin), Gyaur-Kala, and other fortresses. It is likely, if we were to study and map all such districts, that we should uncover appreciable reserves of land the utilization of which would be promising.

Other important man-made indicators are traces of irrigation systems. They may be component parts in oasis relic assemblages, but in a number of places they are independent, where they have very great indicator value, since they point to areas in which irrigation may be carried out even at the present time. The value of these features as an indicator is increased still more by the ease of recognition on air photos. Anyone interested in the details of this problem may acquaint himself with them from the works of Pel't (1951) and Andrianov (1967). We note here merely that, as in the preceding example, there is great value in thorough examination of air photos for discovering areas with characteristic patterns that indicate ancient irrigation systems. The results of such interpretation by man-made indicators would prove to be of fundamental service in planning reclamation or other improvements, especially on ancient alluvial–deltaic plains of the Dzhana-Dar'ya and Kunya-Dar'ya, and also in regions adjacent to Sarykamysh and Assake-Audan, where the land of ancient irrigation is very widespread.

Graveyards and large mausoleums are also easily recognizable both at the site and on air photos. The indicator value of these is somewhat limited, however, since their distribution sometimes depends on causes having no connection with the natural environment (cemeteries appear not only where people live for a more or less protracted interval but also at sites where large battles have occurred between nomadic peoples, at sites considered sacred, at the intersections of caravan routes).

We think that the principal value of cemeteries and mausoleums as man-made indicators in deserts and steppes lies in the fact that they are the most stable, long-preserved relics of oases, features that can be recognized after all other products of man's effort have been in great measure destroyed, leveled with the surrounding surface and therefore ceasing to be physiognomic. Such features may therefore be used only as primary indicators, permitting the investigator to focus his search in a particular region. The final decision concerning the presence of remains of oasis structures may be made, however, only after these other elements are discovered. In combination with other man-made indicators, they may acquire great value. There will be many occasions when the sites of ancient mausoleums and cemeteries may be considered by investigators to be the most notable remains of abandoned oases. For example, some tens of kilometers north of Chelkar occurs the well-known Bolgasyn Mausoleum, described many times in the literature (Matov, 1897; Popov, 1906; Kastan'e, 1908). It stands in a district where traces of irrigation ditches are visible, leading from a nearby lake to the plain adjacent to the mausoleum, and where there are remains of adobe buildings, partially covered by sand. All this is gradually disappearing. Matov, who visited Bolgasyn in 1897, saw the faintly visible ancient irrigation system, but in 1907 it was almost invisible. During an examination of Bolgasyn in 1964, we were unable to detect traces of irrigation, although the general plan of the district made it convenient for irrigation either from the lake or from a nearby stream. This stream locally has a straightened channel and traces of slumping along the banks, so that some features of the ancient oases are still recognizable. The mausoleum and adjoining area are on the surface of an ancient stream terrace with a soil of strongly leached sandy loam, suitable for utilization. The position of the mausoleum is now a unique feature by which one may recognize this abandoned oasis. East of Bolgasyn, between it and Irgiz, in the region of Taumal-Kul', Matov also saw remains of irrigation ditches and dams, and it thus appears possible that the Bolgasyn oasis is merely part of a once much larger system lying west of Irgiz. The same author relates stories of old inhabitants concerning the existence at Bolgasyn in older times of a rather large settlement and a birch forest.

Another example of discovering remains of an oasis by a large mausoleum applies to the western margin of the Malye Barsuki sands. Here, at the boundary of the sands with surrounding loam plain, the Kusaman Mausoleum sits on the top of a hill (Fig. 2), clearly visible from long distances, both in surface and aerial surveys. At our first hasty visit to this mausoleum we did not notice any traces of any early settlement. On closer examination, however, at a second visit, we discovered traces of ancient irrigation, clearly made by damming of takyrs along the margin of the sand massif and by means of a large group of wells occurring here. It is worth attention that this district proved to be so convenient for utilization that a state farm and village have now been established near Kusaman Mausoleum.

In similar fashion it was possible to discover remains of an oasis in the Kyzdar-Shikkan sands between the Ust-Urt Plateau and Chagrai. Here, on a similar hill, occurs a large cemetery and a central mausoleum, under which, according to legend, the Kazakh horseman Kaltybai is buried. The presence of a large cemetery and also some fleeting remarks in the literature concerning fertility of the soil here, the high quality of pasturage, compelled us to examine the area carefully, and this led to discovery of the remains of irrigation ditches, a dam, a protected water reservoir, and a number of traces of embankments. The source of irrigation water was probably a large subsand lens beneath the Kyzdar-Shikkan massif (Abrosimov et al., 1966). This water is so abundant that in some years the discharge zone of the lens rises on the southern margin of the sands to give temporary lakes (we observed the formation of such a lake in 1966).

At the edge of the Kashkarata sand massif, near Kyzdar-Shikkan, there formerly existed an underground irrigation system, as mentioned in the literature (Samokhvalov, 1956). It has been impossible to find traces of this, however. It was probably on the northwestern margin of the sands and supplied water from the discharge zone of a large subsand lense that exists here. The presence of a small oasis in this place is indicated by the faintly noticeable ruins of buildings, traces of embankments adjoining the sands, and the large Sultan-Kazgan cemetery not far from the sands.

There may clearly be some indicator value in so-called devotional (prayer) sites, i.e., areas of several meters (of the order of 20-40) across, bordered by a series of stones (Karutts, 1911).

Fig. 2. Kusaman Mausoleum (photograph by E. Nukhimovskii).

A. Yanushkevich, who visited the Urdzhar region in 1946, pointed up the concurrence of burial and devotional sites with oases. He noted the abundance of graves on all elevated points between tracts that had been sowed at this oasis (Yanushkevich, 1966). He also mentioned the presence of ruins of defense installations, such as small forts, at places where nomadic people set up winter quarters. He also noted the emplacement of octagonal stone figures at places of winter stations of the Khan's tents.

Ruins of forts and fortifications are very similar in their indicator significance to cemeteries and mausoleums. Like them, they are always found in districts having land suitable for utilization, and commonly they have significance purely from the viewpoint of sentinel activity or defense. We may note many examples, however, of fortresses the remains of which represent the last element of a large relict oasis to be preserved from complete disappearance. Therefore, without having quite definite direct indicator significance, defense works may be used successfully in combination with other less physiognomic man-made features of a locality.

Relic oases may have great value for discovering land suitable for small-oasis farming and for organization of a base for cattle raising. Use of the experience of people who have farmed or ranched in the desert permits us to make more reliable and purposeful selection of sites to become centers of utilization and helps to avoid a number of mistakes that frequently arise in solving these problems.

Traces of winter quarters for nomadic peoples ("auls," in Kazakhstan called "kstaus") are rather similar to relic oasis assemblages. The position of the winter quarters or auls was usually related to the presence of steady water supply, the presence of hay fields, and, commonly, the possibility of sowing crops (Rudenko, 1927). Characteristic features of an ancient "kstau" are remains of dwellings of crude bricks and corrals, which had different walls: of brushwood, giant-reed chaff, raw mud, and, lastly, of blocks of undressed stone. The preservation quality of "kstau" traces is very high. In the Malye Barsuki we saw winter quarters abandoned, according to information obtained from local inquiries, 50 years ago, and yet the walls in them were still entirely preserved.

A characteristic feature of earlier "kstaus" is the low earthen ridges which, according to the observations of Sokolov (1908), surround meadow grass hay fields and places where stacks were placed; or the feature may be a series of ditches or moats, used for the same purpose (Bukeikhan, 1927). Walls of corrals are poorly preserved (except where they were made of stones), but the thick layer of dung that formed in the corral exists much longer, and by it we may recognize the site of the corral even when nothing of the walls remains. The central parts of corrals that are covered with dung are clearly visible on air photos as clear, very dark, almost black segments that are round, square, or elongate-rectangular in form. With time, different nitrophilic forms begin to grow in the layer of dung — members of the genera Atriplex, Chenopodium, Sueada — and Peganum harmala L. also grows abundantly. Specimens of these same nitrophilic plants usually grow at the walls of the corrals and in yards. Since crude bricks were used in construction of the winter quarters, bricks made from saliniferous clay, various halophytes also commonly appear in an ancient "kstau," but are absent from the areas surrounding the ruins of the occupied area. These mixed groups of nitrophilic and halophytic plants, produced indirectly by the influence of man, are also very stable man-made indicators, pointing to ancient stations even when almost all other elements have disappeared.

At the present time there is a certain interest in joint utilization of pasture land in connection with the study of the system of abandonment in the past, since some rational elements in this may be found applicable to modern cattle raising in desert regions. In this it becomes necessary to trace the network of earlier cattle trails and movement of nomad caravans. In some cases ancient caravan routes have been preserved and are even utilized for present-day transport. But many of the ancient tracks have geen abandoned and no traces of them remain. In such cases it is also possible to use man-made indicators. Primarily these will be various structures built long ago along the route. Here we should include the well-known "obos" of Mongolia (Pond, 1962) and also many of the barrows (kurgans) of the south Russian steppes. Markov (1901), in his description of the Savvino and Izyum "sakmas" (a "sakma" is an ancient military route of nomads campaigning against the Russ, the road of ancient invasion) mentioned that in traveling over the "sakmas," barrows (kurgans) are "seen to the right and to the left" (Markov, 1901, p. 46). On some of these kurgans, sitting and standing figures (stone sculpture) have been preserved. The connection of the ancient roads with the kurgans was rather clearly traced by us to western Ust-Urt. Here, at an ancient caravan route near the northern branch of the "Ancient Nogai Road" (Grigor'ev, 1861), kurgans have been preserved (Uyuk and others). There are very many kurgans also in the region between the Sam sands and the western break of ths Ust-Urt Plateau. Stone statues have been found on some of these. Alekseev (1963) found figures on Kozybai-Dyurt-Kul' Hill that represented a man with a quiver over his shoulder and a woman with a necklace about her neck. These barrows are clearly related to a road now gone, connecting the Sam sands with the sea coast. This is highly probable in view of the fact that the ancient city Shom is placed by archeologists and historians in the vicinity of Sam (Bartol'd, 1902), and this city surely had some connection with the shore of the Caspian.

An important indicator of ancient roads is vegetation strongly altered by man. Markov cites correspondence of the Valuiki Governor Koltovskii to the Russian Tsar Alexei Mikhailovich, containing a dispatch concerning burning the grass "on the Tatars' sakmas," in order that, in front of that "Tatar encampment, there be no resting place" (Markov, 1901, p. 25). Even one such burning should cause some difference in vegetation in a belt along the road as compared with the surrounding steppe. Regardless of the burning, a strong imprint on the road was made by trampling of the native plant communities and their replacement by pasture weeds. In Kazakhstan the most common indicator of ancient caravan routes is a pure growth of Ceratocarpus, the abundance of which has been noted along highways and roads by many.

In calling the investigator's attention to man-made indicators, we should point out clearly that we do not set this type of indication against others (geobotanical, geomorphic, landscape). but believe that indication by means of man-made features is merely one of the weakly developed links in landscape-indicator investigation. In this we are in complete agreement with the trend of Meier's work (Meier and Nefedov, 1962), in which man-made indicators are considered but one of the varieties of landscape indicators in the broad sense of the word. The use of man-made indicators should not be a special independent objective of one's work but rather a link in the entire complex of landscape-indicator investigations.

The method of using man-made indicators has been worked on but slightly. The simplest and easiest way of using them (as applied to our personal research) is as follows. The first stage is an examination and interpretation of air photos. At this stage we give special attention to the different elements of the air-photo image, which may be readily distinguished by their outlines from the general pattern of the landscape. It is during these first steps of interpretation that the features of man-made indicators appear with special force, involving the fact that traces of man's activity are most notable against a background of some locality which on the whole remained unchanged by man. Rectilinearity, right angles, perfection of geometric forms – all these are important indicators when interpreting man-made elements. A rather common source of error is the similarity with man-made elements of some patterns associated with tectonics of a region. Clearly visible fault lines may be taken for paths, roads, remains of fences, and so forth. The optimum scale for interpreting man-made elements on air photos is 1:10,000 or larger. In the absence of air photos, one should use large-scale maps.

The second stage is a reconnaissance flight over the region with attendant visual observations from the air. At the time of this flight one should pay special attention not only to remains of various structures, but also to the accompanying contaminated vegetation, which clearly emphasizes the traces of man's habitation.

Growths of Peganum harmala L. are especially well marked from the air: dark green in summer, having a characteristic golden orange color in autumn. Observations from the air make it possible to introduce a number of corrections in the results of preliminary interpretation.

If organization of the work permits, the plane flight may be combined directly with surface field description of the indicators. For this purpose a landing strip is marked out, and the segment is then described by ordinary methods of landscape investigation. If the indicator points to traces of agricultural utilization of a given district in the past, it is then necessary to describe the soil and to collect samples for analysis. If the indicator is of pasture land, special attention should be given to the search for sources of water supply for the livestock that gathered here. However, such investigations are most commonly not made at the time of landing, but during special surface traverses made after the reconnaissance flight.

The concluding stage is composite interpretation of the data obtained. Here, very much depends on the experience of the investigator, on his knowledge of local population in the past. The problem of this stage is to combine the information obtained and to compare it with that obtained from other types of landscape-indicator investigations.

In combining different types of man-made indicators, evaluating their age and strictly differentiating indicators of agricultural and grazing utilization, it is possible to prepare a scheme for their distribution in definite regions. Such schemes may have value for preparing schemes of reclamation and irrigation of the land, thus facilitating the development of these measures.

LITERATURE CITED

Abrosimov, I. K., Vostokova, E. A., and Viktorov, S. V., "Landscapes of sand massifs of Kashkar-Ata and Kyzdar-Shikkan and their significance for water indication," Izv. Vses. Geogr. Obshch., Vol. 98, No. 4 (1966).

Alekseev, "Topographic description of the northern part of Ust-Urt and the adjacent region to the Émba River, prepared by the Topographic Corps under Sublieutenant Alekseev," in: The First Russian Scientific Investigations of Ust-Urt, Izd. AN SSSR, Moscow (1963).

Andrianov, B. V., Mapping Ancient Irrigation Systems on the Basis of Air Photos (in Connection with Potential Utilization of the Land), Materials of the Moscow Branch of the Geographical Society, SSSR (Materialy Mosk. Fil. Geogr. Obshch. SSSR), Aerometody, No. 1 (1967).

Bartol'd, V., "Information on the Aral Sea and the lower reaches of the Amu-Dar'ya from earliest times to the 17th century," Izv. Turkest. Otdela Geogr. Obshch., Vol. 4, Scientific Results of the Aral Expedition (Nauchn. Resul'taty Aral'skoi Éksped.), No. 2 (1902).

Battersby, P., Richard Corfield of Somaliland, London (1914).

Bukeikhan, A. N., Kazakhs of the Adaevka District, Materials of the Special Committee on Investigations of the Union and Autonomous Republics (Materialy Osobogo Kom-ta po Issled. Soyuznykh i Avtonomnykh Respublik), No. 3, Kazakhstan Series, Leningrad (1927).

Grigor'ev, V. V., Description of the Khiva Khanate and the Road to It from the Saraichik Fortress, Notes of the Russian Geographical Society (Zap. Russk. Geogr. Obshch.), Book 2 (1861).

Jardin, D., The Mad Mullah of Somaliland, London (1923).

Karutts, R., Among the Kirghiz and Turkmen, St. Petersburg (1911).

Kastan'e, I., "Ruins of Bolgasyn and the Chelkar Steppe," Trudy Orenburgskoi Uchenoi Arkhivn. Comissii, No. 19 (1908).

Lewis, J. M., The Modern History of Somaliland, London (1965).

Macfadyen, W. A., "Taleh," Geogr. J., Vol. 78, No. 2 (1931).

Markov, E., "Along the Savvino sakma," Zhivopisnaya Rossiya, Vol. 1, Nos. 1-4 (1901).

Matov, A., "The ruins of Bolgasyn," Turgai Newspaper (Turgaisk. Gazeta), No. 124-125 (1897).

Meier, G. Ya., and Nefedov, K. E., "Interpretation of groundwater of typical landscapes in Turkmenia on air photos," in: Application of Aerial Methods to the Study of Groundwater, Izd. AN SSSR, Moscow (1962).

Nikitin, S. N., Account of the Expedition of 1892 in the Trans-Ural Steppe of the Ural Region and Ust-Urt (with Appendix of Sketch Maps and 6 Tables of Profiles), St. Petersburg (1893).

Pel't, N. N., "Ancient irrigated lands of the Dzhana-Dar'ya ancient alluvial plain," Izv. Vses. Geogr. Obshch., No. 3 (1951).

Pond, A., The Desert World, New York (1962).

Popov, A. V., "Some comments on the archeology of the Turgai and Ural regions," Trudy Orenburgskoi Uchenoi Arkhivn. Comissii (1906).

Rudenko, S., "Outlines of living conditions of Kazakhs in the basin of the Uil and Sagyz Rivers," in: Materials of the Special Committee on Investigations of Union and Autonomous Republics (Materialy Osobogo Kom-ta po Issled. Soyuznykh i Avtonomnykh Respublik), No. 3, Seriya Kazakhstanskaya (1927).

Samokhvalov, A. M., "Irrigation and water supply of Kazakhstan deserts for development of cattle raising," Trudy Inst. Vodn. i Lesn. Khoz-va Kazakhst. Fil. VASKhNIL, Vol. 1 (1956).

Sokolov, D. N., "A journey across the steppes," Trudy Orenburgskoi Uchenoi Arkhivn. Komissii, No. 19 (1908).

Yanushkevich, A., Journals and Letters of a Traveler across the Khazakh Steppes, Kazakhstan, Alma-Ata (1966).